Nick

The F......g
Ludlow?

23.6.20

Animal Disease and Human Trauma

Also by Maggie Mort

BUILDING THE TRIDENT NETWORK
A Study of the Enrolment of People, Knowledge and Machines

Animal Disease and Human Trauma
Emotional Geographies of Disaster

Ian Convery
University of Cumbria, UK

Maggie Mort
Lancaster University, UK

Josephine Baxter
Lancaster University, UK

Cathy Bailey
National University of Ireland Galway, Ireland

Preface
Kai Erikson
Yale University, USA

© Ian Convery, Maggie Mort, Josephine Baxter and Cathy Bailey 2008
Preface © Kai Erikson 2008

All rights reserved. No reproduction, copy or transmission of this publication may be made without written permission.

No portion of this publication may be reproduced, copied or transmitted save with written permission or in accordance with the provisions of the Copyright, Designs and Patents Act 1988, or under the terms of any licence permitting limited copying issued by the Copyright Licensing Agency, Saffron House, 6-10 Kirby Street, London EC1N 8TS.

Any person who does any unauthorized act in relation to this publication may be liable to criminal prosecution and civil claims for damages.

The author has asserted his right to be identified as the author of this work in accordance with the Copyright, Designs and Patents Act 1988.

First published 2008 by
PALGRAVE MACMILLAN

Palgrave Macmillan in the UK is an imprint of Macmillan Publishers Limited, registered in England, company number 785998, of Houndmills, Basingstoke, Hampshire RG21 6XS.

Palgrave Macmillan in the US is a division of St Martin's Press LLC, 175 Fifth Avenue, New York, NY 10010.

Palgrave Macmillan is the global academic imprint of the above companies and has companies and representatives throughout the world.

Palgrave® and Macmillan® are registered trademarks in the United States, the United Kingdom, Europe and other countries.

ISBN-13: 978–0–230–50697–8   hardback
ISBN-10: 0–230–50697–6   hardback

This book is printed on paper suitable for recycling and made from fully managed and sustained forest sources. Logging, pulping and manufacturing processes are expected to conform to the environmental regulations of the country of origin.

A catalogue record for this book is available from the British Library.

Library of Congress Cataloging-in-Publication Data

   Animal disease and human trauma : emotional geographies of disaster / Ian Convery ... [et al.] ; foreword, Kai Erikson.
     p. ; cm.
   Includes bibliographical references and index.
   ISBN-13: 978-0–230–50697–8 (hardback : alk. paper)
   ISBN-10: 0–230–50697–6 (hardback : alk. paper)
     1. Psychic trauma—England. 2. Foot-and-mouth disease—England— History. 3. Epidemics—England—History.
   [DNLM: 1. Stress Disorders, Traumatic–England. 2. Disasters–England. 3. Disease Outbreaks–history–England. 4. Disease Outbreaks– veterinary–England. 5. Foot-and-Mouth Disease–history–England. 6. History, 21st Century–England. 7. Stress Disorders, Traumatic–history–England. WM 172 A598 2008] I. Convery, Ian, 1965–
   RC552.T7A55 2008
   362.196'8521–dc22
                                               2008016152

10 9 8 7 6 5 4 3 2 1
17 16 15 14 13 12 11 10 09 08

Printed and bound in Great Britain by
CPI Antony Rowe, Chippenham and Eastbourne

Contents

List of Figures and Table		vi
Acknowledgements		vii
Preface by Kai Erikson		viii
Introduction		1
Chapter 1	Disasters	5
Chapter 2	Global to Local: The Case of Foot and Mouth Disease in Cumbria	13
Chapter 3	Of Humans and Animals	41
Chapter 4	Things/Materials	61
Chapter 5	Trauma and Traumatic Experience	89
Chapter 6	Working on the Frontline	114
Chapter 7	Exploring the Lifescape	132
Chapter 8	Reconfiguring Disasters	151
Appendix I	The Compulsory Cull	155
Notes		157
Bibliography		167
Index		182

List of Figures and Table

Figures

1.1	Economic and human impacts of disasters, 1973–2002	6
2.1	The location of Cumbria (shaded area) within the UK	24
2.2	The distribution of FMD cases in Cumbria 2001	25
2.3	Distribution of FMD cases in the UK 2001	26
2.4	'No entry' Cumbria 2001	26
2.5	Ethnoplot of self-reported quality of life and health, together with diary data 'mapped' from December 2001 to May 2003	37
3.1	The Tree of Everlasting Life	42
3.2	Complexities of livestock–farmer relationships: the case of Holstein–Friesian dairy cattle	56
3.3	A life's work	58
4.1	'Keep Out', Cumbria 2001	65
4.2	No entry signs on footpath gate, Surrey 2007	66
4.3	Westmorland Show 2002	67
4.4	Licence to enter or leave infected premises	70
4.5	This footpath is closed, Cumbria 2001	71
4.6	'A Notice' access restrictions	71
4.7	Farm cull	78
4.8	MAFF letter to infected premises	80
4.9	Altered sense of place	80
4.10	Milburn School 'sheep'	82
5.1	'Shutdown'	94
7.1	Altered lifescapes during FMD	143
7.2	Disrupted Lifescape: Eddie	144
7.3	Disrupted Lifescape: Carole	147
7.4	Disrupted Lifescape: Jean	148

Table

2.1	Occupational and residential groups included in the rural citizen panel, and their study data	35

Acknowledgements

We owe special thanks to the people who endured the 2001 FMD disaster in Cumbria and to the panel who gave us their time in interviews and in writing more than 3,000 diaries about their everyday experiences. Without them this book could not have been written.

We are also grateful to the following for their generous assistance: Peter Tiplady, Sue Allan, Brian Armstrong, David Black, Kate Braithwaite, Clare Edwards, Jackie Fisher, Stephen Greenhalgh, John Houlihan, Dee Howkins, Brian Lightowler, Kath Mason, John Pinder, Geoff Sharpe, Veronica Waller, Rob Walker, John Wilkinson, Jennifer Wilson, Cathy Connolly, Sue Wrennall, Kim Riley and Sheryl Coultas.

We wish to thank the following who have informed our work: Emily Bishop, Sue Black, Helen Clish, Mark Duerden, Tony Gatrell, Caz Graham, Alison Gurkey, Anne Hopper, Sandra Hudson, Andrew Humphries, Guy Hurst; David Large, Patricia Macdonald, Emma O'Dwyer, Sol Picciotto, Miodrag Popovic, Kath Smart, Judy Stalker, Diane Steadman, Alan Stones, Brian Strong, Marianne Teasdale, Miles Thomas, Nigel Thompson, Jean Turnbull, Nick Utting. We thank also the members of HM Forces, Cumbria Constabulary, DEFRA, BusinessLink, CADAS, Women's Refuge, CARL, and private individuals who granted us interviews.

We are indebted to Cumbria Community Foundation, Rural Regeneration Cumbria, The Hadfield Trust, and the Friends' Fund of Lancaster University for their support of the 2003 Voices of Experience Conference, and we thank Ron Bailey, Anomie Baxter, Border Reiver Activities, Sue Flowers, Alison Marrs, Nick May, James Popps, Eden Artisans and Woolclip for their contribution to the conference. We are grateful to the following who allowed us to learn from and contribute to their deliberations: Phil Thomas and the Secretariat of the Cumbria Foot & Mouth Inquiry; Caroline Lucas MEP, Elze van Oudenaarden and Sjef Coolegem of the European Parliament Temporary Committee on Foot and Mouth, and John Mallinson, Cumbria County Council. Finally Kai Erikson, Yale University, for visiting Cumbria, offering us his advice and sharing his experience of the unseen and enduring impacts of disaster on individuals and communities.

Preface

I first visited Cumbria in the spring of 2003; two and a half years after the foot-and-mouth epidemic of 2001 began its deadly sweep across the countryside. I came at the invitation of the authors of this book, in part so that I could confer with them about the research they were undertaking, and in part so that I could listen to, and then address, a gathering of people at the Racecourse at Carlisle in North Cumbria. The gathering was made up in large part of persons who had been exposed to the Foot and Mouth epidemic and had suffered from its effects. It was entitled 'Voices of Experience,' and its stated purpose was for officials and for specialists of various kinds to hear the testimonies of individuals who knew the disaster at close hand. I was the last speaker on the program.

When I was first invited to contribute to the volume, the authors and I thought it would make good sense for me to convert the comments I made on that occasion into a more formal preface. After all, I had brought several pages of careful notes with me across the Atlantic, and I proposed to put them to work one more time. With that object in mind we arranged for a transcript of those proceedings to be prepared. Once I read through the transcript, however, I realized how often I interrupted the flow of my prepared notes to turn to subjects that had been brought up by the speakers who preceded me on the podium. The talk had become part of an interaction rather than an address that could stand on its own, and when the time came, a year or so later, for me to try to disengage the words I had spoken that morning from the context in which I had spoken them, I was quite dissatisfied with the result. It was the *interaction* that gave those words life and meaning.

The transcript indicates that I ended my comments at the Carlisle Racecourse by saying: 'You have been very generous with me today. When I am finished talking, you will probably applaud. If you do, I will applaud in return. What we heard this morning was both beautiful and horrifying, and it reflects the spirit of a community I have come, in a very short time, to respect greatly. So if I may ...' And the transcript itself ends with the following, set off in brackets: [Kai leads the applause and closes].

I meant that not only as a salute to the audience but to the people of the region more generally. I would like the paragraphs below to remain so. That would be in the spirit of the event. It would also be in the spirit of this book. The chapters that follow, too, are the continuation of an interaction that has been underway since the outbreak of the epidemic. On the one side we have a group of distinguished scholars who know a great deal about foot and mouth disease and about the human face of disasters, but who have spent most of their time over the past several years listening to others – knowing that the true experts of that terrible time are the people who experienced it. On the other side, we have victims of a kind of hell who probably do not have as much confidence in others as they once did, but who clearly have come to trust the wisdom and humanity and compassion of those scholars.

I now think that I should leave the words I spoke then more or less as they are in the transcript. I have edited the remarks lightly, as writers always do, in the hope that I will appear perhaps more articulate and more self-assured than the audience who were there that day will remember me as having been. But I did less editing than I was tempted to because I would rather sound like one of the voices in the conversation than as a voice commenting on it from a place outside it. Here it is:

Dear friends:

Given the nature of our topic today, it may seem absurd for me to begin by saying what a pleasure it is for me to be here today. But it truly is that. Adam introduced the panel of research respondents this morning by suggesting a distinction between individuals who are not very experienced at public speaking and veterans of the process. I suppose I qualify as a veteran in the sense that I have done this many times. But I think everyone here will understand if I ask you to pity the poor veteran who not only has to come to the podium following that extraordinary exhibit of eloquence and courage and dignity from the respondents, but who also has to speak as an 'expert' on matters that Adam and the others know so much better than he does.

Among my pleasures this morning was to be interviewed by the press on several occasions. One of the first reporters I met asked the most direct, and probably the best, question of all: 'what are you doing here?' That gave me a moment's pause. I think I'm here because I'm thought to be an expert on catastrophic events. I'm not entirely certain right now what that means. If an expert on disasters is someone who has visited a number of them, then I qualify. But I hope it is as clear to you as it is to me that everyone in this hall knows more about what happened in Cumbria than I do. I came here to learn from you. What I hope to be able to do in the time allowed is to share with you a sense of what disasters look and feel like in general as distinct from what this particular one looks and feels like to those of you who lived through it.

When I was relatively new to this grim business, I met with a roomful of persons who felt that they had been damaged by a toxic spill that had contaminated their homes. My reason for interviewing them was to get a better idea as to whether I might be helpful to them in a lawsuit they were in the process of mounting. I asked them questions about the spill, how they had reacted to it, and so on. When I was done, I said, professor-like: 'I've asked you many questions. Do you have any of me?' I did not expect a response, in truth, but almost every hand in the room went up, and their questions, though phrased in various ways, were essentially the same: 'Are we unusual? Is there something about what we have just told you that makes us different from other people?' One man even asked: 'Am I out of my mind to be thinking as I am?' I felt very badly, because I had been introduced to them as an 'expert,' and yet had somehow managed to forget one of the few things I felt that I knew about human responses to disaster – that in one way or another, at one level of consciousness or another, virtually all individuals who suffer the after effects of a disaster wonder whether they are reacting in an aberrant way. It can be a lonely, dark, solitary feeling, because people in many parts of the world, certainly yours and mine, are not very good at sharing their innermost fears with friends or neighbors or even kin. We fear that there is something wrong with us, and since we so rarely take the risk of letting each other know of those feelings, we may take a long time to learn that they are common to most (and perhaps to all) of our fellows. We are surprised to learn how widely a vulnerability that we took to be an individual weakness of our own is shared by others. As we heard this morning: 'what happens to people is an absolutely normal reaction to an event that is itself abnormal.' Yet it can take people a long time to learn that lesson.

I would like to spend a moment noting some of the damaged places I have visited and studied. I do this not to impress you with the number of frequent flyer miles I have accumulated over the years but to give you some sense of the range of experience that specialists like myself are drawing on when we talk in a general way about disasters. I would then like to offer four findings that those of us who do research on the topic feel fairly confident of. And I would like, finally, to reflect once again on what we heard this morning. Those places include:

> A coal mining valley in West Virginia where a winter flood, caused by a faulty dam, did a terrifying amount of damage to the people who lived there.
>
> An Ojibwa Indian Reserve in Northwest Ontario where a mercury spill acted to contaminate the waterways along which people had lived and from which they had drawn sustenance since the beginning of what they reckoned to be time, leaving them with what a wise elder called 'a broken culture.'
>
> A nuclear power plant in Pennsylvania called Three Mile Island where a near meltdown occasioned an extraordinary expression of dread on the part of people who lived in its shadows.
>
> A migrant farm worker camp in South Florida where a number of poor persons from Haiti found that the money they had earned working the fields, money their families in Haiti had been counting on for everyday subsistence, had been stolen from them.
>
> Native villages along the coasts of Alaska where the waters from which people had been taking a living for centuries were blackened for a thousand miles by a tremendous oil spill, one that began by threatening the native economy in ways that go beyond easy measuring and ended by threatening the native way of life altogether.
>
> A town in Western Slovenia that was more or less destroyed by tides of civil war that swept across the Croatian countryside and other portions of what was once Yugoslavia.
>
> A remote atoll in Micronesia that was visited half a century ago by an enormous cloud of radioactive fallout as the result of a nuclear test that was conducted by the United States.

This is only a sampling, but it is difficult to imagine a more varied assortment of horrors. A flood; a mercury contamination; a nuclear emergency; an act of larceny; an oil spill; a fearful civil war; a radioactive fallout; and, if we include the roomful of persons I spoke of interviewing a few moments ago, an underground seepage of petrol. Add to that roster the hundreds of other disasters that are described and analyzed in what we social scientists like to call 'the literature,' and that variety would of course be greater yet. Not every specialist in the study of disasters will agree with what I am about to say, but I will nonetheless assert that the following findings are characteristic of so many disasters that we would be advised to approach any particular specimen of the species with those points in mind.

Severe disasters can do a great deal of damage to the tissues of the human mind and to the tissues of the human body. We call this 'trauma.' At the same time, however, a severe enough disaster can also do serious damage to what I might call the 'tissues' of the community of which the victims are a part. I don't know if 'tissues' is a permissible term under the circumstances, but I really do see it that way. I once made a distinction between 'individual trauma' and 'collective trauma' – meaning damage done to the fabric of the larger community that is quite independent of the damage done to individual persons. We are moving into a world of metaphor here – 'tissues,' 'fabrics' – but I think most persons who have experienced a severe disaster will know what I am getting at even if they, like me, are not sure what words are best suited to the case.

By individual trauma I mean a sense of vulnerability, a sense of no longer being in control, a sense of disorientation and loss of meaning, a sense of despair, and, often, a dark, abiding, aching depression. In addition, there is a strong sense that something is pending, about to happen – that the world is now organized in such a way that something ruinous is almost sure to take place. I have heard the expression 'a ticking time-bomb' any number of times elsewhere, but the expression I heard this morning, 'an unexploded bomb' is better yet by far. There were echoes of all those feelings in the testimonies of this morning, and I thought I saw nods of recognition out there in the audience when I was speaking of them just now.

The second finding I would like to mention is that the level of trauma people are likely to experience after a disaster is far more severe if some kind of toxicity is involved. Toxic disasters are considerably more common now than they have been in the past, for obvious reasons, and they are greatly feared. Nor is that hard to understand. Most of us tend to see 'dangers' as something out there in the world around us trying to force their way through the various defences we build around ourselves. But toxins slink silently and treacherously into the interior of the body, and do their dread work from the inside out. They are the very epitome of perfidy and stealth.

The third finding is that the level of trauma people are likely to experience after a disaster has a good deal to do with whether they regard the event as a product of natural forces or as a product of human miscalculation or even malice. It would seem obvious on its face that if other human beings prove to have been responsible for, say, contaminating your home – all the more so if they are a part of the same social world you live in – you will be angry at them and perhaps even begin to wonder whether the rest of humankind can be relied upon in moments of urgency. What is not so obvious, but just as important, is that the sense of injury one experiences as the result of a disaster can be a good deal sharper if survivors come to feel that the people around them respond with what seems to be a kind of indifference or denial; if they act as if nothing of consequence had happened, or if they imply, by the way they behave or by the words they utter, that the survivors are over-reacting, a bit hysterical, or in some other way not responding in the way they ought to. The sense that others do not fully appreciate what is happening to one can easily ripen into a feeling of having been let down, of having been left out of the human community, even of having been betrayed. That is a very common outcome, too, particularly when companies responsible for an emergency are called upon to defend themselves, and when government officials asked to do something about an emergency are criticized for their inability to be of more help. Particularly when the threat of legal action hangs in the air, companies have little choice but to try to belittle the claims of the plaintiffs suing them, and it is understandable, if lamentable, when officials, frustrated by their failures, come to see the fault as residing in the victims. That is a very human story.

That brings me to the fourth finding – an extension, really, of the third. The sense of betrayal I spoke of a moment ago can, in its turn, easily ripen into a suspicion, lodged deep in the heart, that the rules governing the social world, and perhaps even the rules governing the natural world, have been suspended for at least the immediate time being. I once called this 'the traumatic worldview.' I meant by that term to suggest a feeling shared by individuals who have experienced severe crises that they have become different as a result, that they now know things that other people are not aware of, that they now look out at the world through a differently shaped lens than they did before. It is a recognition that the world is not regulated so much by order and continuity, as our parents and teachers told us to be the case, but by a kind of raw chance and maybe even a kind of raw malice, that lurks out there near the edges of things and operates in ways of which we have no knowledge and over which we have no control.

I may not have had courage enough to read the next quotation if someone had not mentioned the holocaust this morning, and I hope everyone will understand that I am not suggesting anything even approaching equivalency by alluding to the one in the context of the other. One holocaust survivor, however, speaking fifty years after his internment, put what I have been talking about in chilling perspective:

> It's more like a view of the world, a total world view of extreme pessimism, of really knowing the truth about people, about human nature, about death, of really knowing the truth in the way other people do not know it.

To live through a serious disaster, then, as many persons here have reason to know, is to feel as though one sees the world anew, as though one can discern the dark and the bleak and the gloomy in human life with eyes made sharper by adversity.

Those are the four findings I hoped to share with you. Having done so, I propose now to turn back to comments made this morning and discuss them in that context.

The first point I raised had to do with collective trauma. Someone in the morning panel noted that she had seen communities in pain before, but that this one appeared to be truly hurting. That would be a reflection of collective trauma in one of the senses I have in mind. But another form of collective trauma, one that may also prove to be characteristic of this countryside, occurs when fault-lines open up across the community that the people who live there were not aware of. I think I have been hearing people say here that there are divisions in the community now marking off those who were culled from those who were not. I did not even know when I flew here the day before yesterday what the word 'culled' meant. But I have heard similar comments elsewhere, and colleagues of mine, who have seen that process at work, speak of 'corrosive' communities.

Yet another form of collective trauma – one I think I have heard people refer to here – occurs when a line is drawn around the affected community and sets it off from the surrounding countryside. It is almost as though the persons who live within that bounded territory feel that they have become a different kind of person from those who live outside it. That is partly because the persons who live within the marked off space have shared a rare and terrible experience. But it is also because the persons who live outside the marked-off space are anxious to *view themselves* as different. They hope, without even knowing they are doing so, that to be of a different kind, to live under different stars, to share a different history, will provide something like immunity to the disaster that visited their fellows. It makes them feel safer.

I think it was Adam who referred this morning to 'the feeling that it is partly our fault.' I've heard that many times. Guilt is a hard human emotion. It is natural for us to ransack our memories for reasons to suppose that *we* were somehow responsible for what happened to us. If nothing else, it at least serves as an explanation for the inexplicable. But those feelings are often reinforced by individuals on the other side of the invisible dividing line, because if they can persuade themselves that the victims brought troubles upon *themselves*, it reduces the chances that the same will happen to them. It is like a magic talisman.

The second finding I drew attention to had to do with toxicity, and I brought the subject up at least in part because I wanted to hear what you think about the following. I have for a long time thought that 'toxic' should not be defined chemically for purposes such as the epidemic we are discussing today. If one used the term 'toxic' to refer to the *reactions a substance evokes* rather than *the physical properties it contains,* we would be in a position to point out that viruses go to work on individuals, human communities, and even animal herds in essentially the same way as radioactivity, dioxin, or any other form of toxicity does. It moves inside stealthily and without warning, setting up a base camp, as it were, in that hidden interior, and then destroys living tissue from the inside out. It acts as poison does.

The third and fourth findings I mentioned a moment ago have to do with feelings of distrust people commonly develop after an encounter with disaster. The main point to be made is that those feelings often radiate outward and come to bear on ever wider circles of social life. It can begin with doubts that reach out to certain of one's neighbours, local officials, or other figures in the immediate community, but it can then expand to include human institutions more generally, or even to the ways of the natural world or the ways of heaven.

I propose to close with a short list of things that people who were not involved in the furore of attending a disaster will seldom understand about those who were. One of them has to do with the grief victims often feel over the loss of familiar things in their universe that seem to others as not deserving so deep an emotion. I've talked to many victims who feel genuinely distraught because of the loss of a garden that they had brought to life over the course of many years, or the loss of an addition to one's home that had been created in slow instalments on Sunday afternoons. I have seen this feeling most poignantly when victims try to explain to others what it is like to lose something, the value of which can be easily replaced, in which they have invested a part of themselves. It is hard to explain to oneself, virtually impossible to explain to others. The law sees human beings as entities who exist within an outer skin. That is the membrane that separates us from everything else. But the fact is that most of us develop a sense of who we are and where our selves begin and end that includes other things we have made ourselves or in which we have an emotional investment. They are extensions of our persons. And when those things are wrenched from us, our sense of self can be greatly reduced. A part of us has been taken away.

You all know where I am going with this thought. It will be very hard for others to fully appreciate what it means to grieve over the death of a farm animal, particularly one that was scheduled for slaughter in any event. I will not go further with this, since everyone in this hall knows more about the topic than I ever will, but I can at least bring you the news that people elsewhere who have suffered similar losses have had the same experience. They, at least, will know what so many of you are going through. It may take them a longer time to understand that the loss of heads of cattle is a different thing than the loss of a herd, an organic whole, in its own way a community. I did not know enough to utter that sentence myself but two days ago.

There is yet another thing you can count on most other people to not understand – that those who have experienced a disaster up close are likely to measure the passing of time differently than those who have not. Those other people will not understand what you mean when you locate an event by saying 'well, that was *before*,' as if the calendar itself was now built around the disaster. Nor will they understand how hard it is for you to place the disaster in the past.

Most people out there look at a disaster in the same way they do any other event that punctuates the passing of time. For them a disaster has a beginning. It has a middle. And it has an end. It obeys Aristotle's rules of plot. Whether we are talking about a flood or a fire or an epidemic, the event has a beginning and middle, and it ends when the flood waters abate, the fire cools, or the epidemic is declared to be over. At that instant, it officially becomes an event in the past, and if conversation continues about its effects or something of the sort, we will call it 'aftermath.' It is not a part of the disaster, but its sequelae.

But that is not how a disaster is usually experienced by those who were involved in it. For them they remain a part of everyday life, the disaster has not become a thing of the past simply because the flood waters have dissipated or because the flames have been extinguished or because officials close their briefcases and declare the epidemic to be over. Disasters linger in the mind. They appear in the thoughts of the day and in the dreams of night. They are experienced as an ongoing event, a continuing calamity, long after others have turned the page to a new day, and it will remain so until the minds of those who lived through it reach a certain peace. That can be a very difficult thing to explain to people who still calculate time in a way you no longer can. You have been very generous with me today ...

And that is how my remarks came to a close at the Carlisle Racecourse on the 14th of October, 2003. I would like to close this preface by saying one word further. The authors of this volume speak of a remarkable and powerful documentary film by Spike Lee called *When the Levees Broke*, a report on what Hurricane Katrina did to the city of New Orleans and its people. What gives the film its power, they suggest, is that the voices heard throughout are not those of outside experts speaking of something that happened to others, but those of the victims speaking of what happened to themselves. The authors are quite right to make that connection, because among the many merits of the book you are about to read is their understanding that the minds of the persons who experienced the epidemic is where it resides, where it has its true meaning. In being so sensitive to that truth, the authors have written a very special and important book, and I would like to close these remarks with a salute to them.

Kai Erikson
Yale University

Introduction

A great number of books and papers have been written about disasters. In our book we wanted to join with a minority of accounts which offer ways of thinking about disasters in non-linear, non-prescriptive and humanistic ways. What we mean by this is that many disasters are approached as if they have a clear beginning, middle and end, an approach often found, for example, in disaster management. But the experience of being in a disaster often isn't like this – it is messy, it may resonate strangely with previous events and experiences which don't 'count' as disastrous, and it may only be generations later when a disaster reaches its grim potential. And for many victims or survivors, what makes events so harrowing is a sense that the past, the present and the future are all implicated or affected somehow by what has happened. By 'non-prescriptive' we mean that we do not take a particular (professionalized) approach to disasters and there is no fixed definition of disasters that readers will find clearly in this book. We do not begin our book with a set of assumptions about one event, or groups or set of materials which are any more important than any other. And we do not offer blue prints or protocols for managing disasters, except for one very simple point (and this is the humanistic bit) we have come to believe that it is those who are 'inside' disasters who should define what such events and processes mean, who should be given space and time to articulate what they need and what helps them. Yet this is often that last thing that is considered in the rush to manage the latest disaster. Of course all disasters are different, they have important specificities and we are not saying that immediate relief efforts are wrong. But we are saying that many disasters take time to unfold and to understand, and that the consequences of not listening, not being alongside those who are involved, can lead to a wholesale loss of faith in the possibility of recovery, in recovering the 'social'.

Because we collected a great number of stories in our work with people who had lived through one disaster: the 2001 UK Foot & Mouth Disease (FMD) epidemic, inevitably we draw on these heavily in this book. But, reflecting the concerns of many of those who told us their stories that our work should try to make a difference, we place these narratives within a wider consideration of disasters and try to rethink some of the boundaries that get made by policy makers, managers and commentators. Boundaries such as the local and the global scale; the 'natural' or the 'technological'. Of course, our sense of these boundaries can also suddenly shift when disaster strikes, challenging long-standing senses of place, contact, culture, identity and humanity. But we hope to question these boundaries by exploring those points at which they appear most interlinked – and this is what we call the emotional landscapes of disaster.

Our work on FMD, and subsequent approach in this book, has been inspired and influenced by three main sources. First, the work of the sociologist of disasters, Kai Erikson, whose humane and accessible writing is a touchstone for all who wish to understand the unseen and enduring impacts of disaster on the lives of individuals and communities. That disasters create 'insiders and outsiders', a kind of collective identity, is discussed by Erikson (1994) from his work in many different disaster settings. In exploring trauma, he draws an analogy with the body by suggesting that, 'when the community is devastated...one can speak of a damaged social organism in almost the same way that one would speak of a damaged body.'

Second, the innovative and insightful approach of Takashi Harada, whose work considers the relations of spaces, materials and the 'social' from his ethnography in a public shelter following the Kobe earthquake in Japan, January 1995. Harada argues that 'individuality makes sense only in a social context...we construct ourselves and we are constructed both "socially" or "individually" in space and with relations with the materials.' When familiar spaces and materials are changed, we are forced to reconsider our social and individual relations.

We also draw on McLean & Johnes' incisive re-telling of the 1966 Aberfan disaster in which 144 people were killed, 116 of whom were children, buried when a coal slag tip slipped down onto the village school. A local chapel was used to house the bodies as they were recovered and was the place where parents came to identify their children. Later villagers asked that the National Coal Board (which was apportioned liability) should fund the demolition and rebuilding of the chapel as they felt they could no longer worship in a place which had such a traumatic association with their loss. This request was initially refused and is given as an example of how those outside of the disaster view such requests as 'irrational', while for insiders it is a normal reaction to an abnormal event.

But above all these influences in shaping this book, important as they have been, we owe most to the people of Cumbria who shared with us their memories, experiences, and daily troubles and triumphs in the wake of the disaster. Their generosity in taking the time to talk with us, write for us and even come back and comment for us on the whole process, has given us a lifetime's material with which to work. This was an enormous privilege and has inspired and driven us to contemplate adding to the already extensive literature on disasters.

Disasters change things. Relationships within and between communities can fracture, realign, and coalesce in new and unexpected ways. Familiar places and materials take on new meanings and identities. During the 2001 FMD crisis, for example, the drive to work, unavoidable for many rural people, took on new meaning when it involved passing pyres of burning animal carcasses prompting the driver to fear that s/he may become a vector of infection simply by following a daily routine. On farms, spaces such as the cow shed became eerily silent once the dairy herd had been slaughtered. What is normal and what is abnormal can shift dramatically. These are themes to which we will return frequently in the following chapters, each of which serves as a lens to view selective aspects of the disaster experience from an 'insider' perspective, drawing on both the insights of particular informants and the interview transcripts and diary material from what is the UK's largest social study of the 2001 FMD epidemic.[1]

Chapter 1 begins the task of defining disaster, which we find to be peculiarly difficult since what counts as 'disastrous' shifts, along with the ever changing forms that catastrophic events are taking and the range of emotional responses to them. In Chapter 2 we offer a brief history of FMD before considering the characteristics of the UK 2001 FMD epidemic. The complex relationship between farmers and livestock forms the context for Chapter 3. In Chapter 4 we consider the role of 'things': personal effects, materials, spaces, iconic objects and how their size and significance changes in disasters. A major theme of our work is explored in Chapter 5, that of individual and community trauma and our aim, along with Erikson and others, to de-pathologise the idea of trauma in the context of disaster. Chapter 6 considers trauma from the perspective of those working on the 'front line' of disasters, and offers some thoughts on how they might best be supported.

Chapter 7 considers the theoretical and 'lived' aspects of lifescape, we attempt to draw much of our work together in this concept, which we have found so useful. As we hope to show, thinking about the lifescape helps to bring together, even if temporarily, what is fractured both in the disaster itself and ironically, often repeated by those agencies which are tasked with recovery and regeneration. Then in Chapter 8 we reiterate our plea for those inside disaster to be seen as the experts, to have their experience acknowledged and beyond all, to teach us what we have so often lost, that the place in which we live is an emotional landscape.

1
Disasters

Disasters

Globally, and if one was counting, it would seem that disasters are on the increase. The United Nations Development Programme (UNDP, 2004) states that existing records, going back to 1900, show a relentless upward movement in the number of disasters and their human and economic impacts. Disasters can lead to widespread loss of life, both directly and indirectly, affect large segments of the population and cause significant environmental damage and large-scale economic and social harm (United Nations Economic Commission for Latin America and the Caribbean (UN/ECLAC), 2003). According to the World Heath Organization Collaborating Centre for Research on the Epidemiology of Disasters (CRED), more disasters were reported for 2000 than in any year over the last decade, affecting some 256 million people. Both CRED[2] and The United Nations International Strategy for Disaster Reduction (UN/ISDR, 2004) reports that the trend during the last three decades shows an increase both in the number of natural disasters and their related costs (Figure 1.1).

Natural disasters (of which more later) represent less than 13% of the UN/ISRD reported deaths from disasters during the period 1900 to 1990, with famine (39.1%) and civil strife (48.6%) far more important causes of mortality (Blaikie *et al.*, 1994).

Of course, this raises the question of what constitutes disaster? At first this might seem a fairly straightforward question, but 'disaster' eludes simple definition,[3] or as Philip Buckle (2007) puts it, defining disaster is never easy and rarely definitive. Indeed, the word *disaster* is so frequently used in everyday talk, from misplaced house keys to major events such as earthquakes and hurricanes, as to be almost meaningless. This ubiquity is

Figure 1.1 Economic and human impacts of disasters, 1973–2002

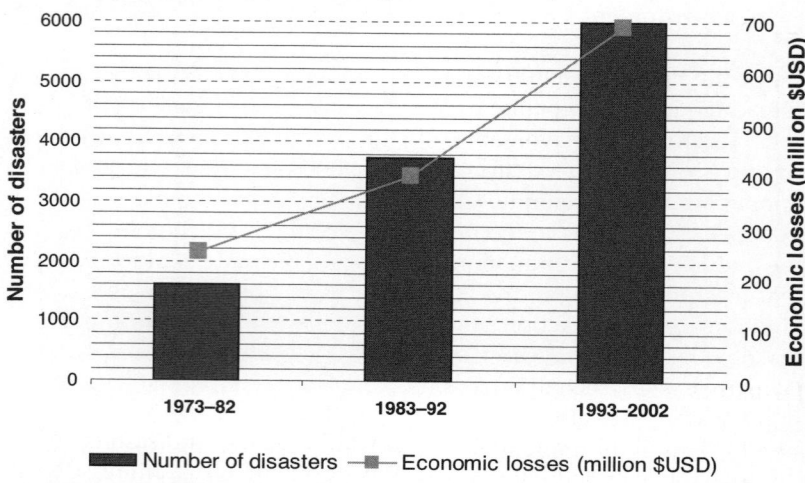

Source: UN/ISDR (2004)

problematic, and as López-Ibor (2005) notes, in academic disciplines it is almost impossible to find an acceptable definition of a disaster. The term originates from the unfavourable aspect of a star, from the French *désastre* or Italian *disastro*, and suggests that when the stars are poorly aligned unfortunate things are likely to happen; the implication is that *disastrous events are outside human control*. Indeed, disasters may still be viewed as 'events from the physical environment...caused by forces which are unfamiliar' (López-Ibor, 2005:2) and frequently unforeseen.

Nesmith (2006:59) writes that the word disaster has many synonyms that add conceptual significance to the term in communicating misery, death, destruction, helplessness, sudden reversal of what is expected and unhappy resolutions to distressing events. She provides a set of defining characteristics:

- Event that disrupts the health of and occurs to a collective unit of a society or community
- The event overwhelms available resources and requires outside assistance for management and mitigation
- Represents tremendous relative human losses
- Negative impact event of natural, financial, technologic, or human origin, for example, armed conflict

- Represents a breakdown in the relationship between humans and the environment

Nesmith's criteria, useful though they are, nevertheless reflect a widespread tendency to separate natural disasters, which have traditionally been regarded as unavoidable and to a large extent beyond human control (i.e. an act of God) from human-induced disasters, which includes human error – for example, Chernobyl – and intentional activity – for example, warfare (Weisæth *et al.*, 2002). We find this dichotomy between the 'natural' and the 'human made' to be problematic, and would argue that regardless of origin, the consequences of a disaster involve a combination of human action and interaction with the environment. As Blaikie *et al.* (1994) state, in their excellent account of the social dimensions of disasters, such disasters are a complex mix of natural hazards and human action (see also Skinner & Mersham, 2002). O'Keefe *et al.* (1998) also highlight the ambiguities in the ways in which disasters are identified, and state that whilst some natural event may act as a trigger for disaster, economic, social and political factors also play an important role. Similarly, Hearns & Deeny (2007) argue that many of the world's disasters are now complex emergencies because they include a multiplicity of problems such as war, ethnic conflict, famine, endemic diseases, and political unrest.

Disasters take many forms and their impact is highly situated rather than absolute. Disasters are culturally, spatially and emotionally specific; a product of the social, political and economic environment, as well as the natural environment, because of the way the lives of different groups of people are structured (Blaikie *et al.*, 1994). Similarly, McLean & Johnes (2000) state that disasters are embedded within specific socio-economic, historical, cultural and chronological contexts, and note that a personal disaster may go unnoticed outside immediate friends and family. Thus whilst some disasters are high impact events, causing great loss of life and trauma; others are low impact and small scale. Most people would rightly acknowledge the 2004 Tsunami which swept across Southern Asia as a disaster, it is likely that less would mention the collapse of the Equitable Life group in 2004. Yet whilst the Tsunami resulted in large scale loss of life and extensive damage to property of some of the poorest and most vulnerable people on Earth, the latter resulted in around 800,000 people losing money, in some cases wiping out life savings and pension plans. To many of these people, the collapse of Equitable Life was a disastrous event in *their lives*. In the same way, the collapse of the UK Christmas hamper firm Farepak in 2006 cost thousands of families (typically on low incomes) their annual savings. This is not to undermine the devastating effects of the Tsunami, but merely to illustrate that events do not need to result in widespread loss of life and destruction to qualify as a disaster, lives can be devastated without death or structural damage.

In this sense, the broader definition of disaster, adopted by many international agencies and NGOs, is surprisingly useful. The World Health Organization (2002) states that a disaster is any event that exceeds the capacity of individuals or communities affected to alleviate their suffering or meet their needs without outside assistance. The United Nations Development Programme (2004) also highlight 'a serious disruption of the functioning of society, causing widespread human, material or environmental losses which exceed the ability of affected society to cope using only its own resources.' The Australian emergency manual series (Emergency Management Australia, 1998) defines a disaster as 'a serious disruption to community life which threatens or causes death or injury in that community and/or damage to property which is beyond the day-to-day capacity of the prescribed statutory authorities and which requires special mobilization and organization of resources other than those normally available to those authorities.' Similarly, Blaikie et al. (1994) state that disasters occur when 'significant number of vulnerable people experience a hazard and suffer severe damage and/or disruption of their livelihood system in such a way that recovery is unlikely without external aid'. Even with external aid, disasters may have significant impacts on the living conditions, economic performance and environmental assets and services of affected areas (UN/ECLAC, 2003). Consequences may be long term and may even irreversibly affect economic and social structures and the environment.

Not only do disasters affect social functioning, they are also the consequence of history, reflecting past events and failures, as with the Aberfan disaster (McLean & Johnes, 2000). Such a context provides a different lens to view disasters; it enables us to see beyond the initial event and view disasters as happenings that entangled the local and the global, the historic and the future, continuity and dramatic change. Disasters are seldom discrete temporal events. Furthermore, such histories allow us to define and shape community social vulnerability (Fortun, 2001:1), and can be used to highlight how, for instance, the Chernobyl nuclear disaster[4] that took place on 26 April 1986 was not just an event, but an ongoing process, 'a process that we are living with and living in' (Rahu, 2003:295). Upland animal movement restrictions in the UK (parts of Scotland, Cumbria and Wales) were imposed due to the fallout from the Ukranian reactor and the subsequent uptake and re-uptake by sheep of radiocaesium in the soil. The underestimation of the extent of this problem by Government scientists has been explored by Wynne (1997). It is a stark illustration of this point about the nature of time in disasters to note that 21 years after the Chernobyl disaster, nine farms in Cumbria are still under restriction (North West Evening Mail, 2007). Similarly, Fortun (2001:354) writes that the Bhopal disaster[5] comes to operate like the cumulative effect of toxics – inhabiting us, reconfiguring us in ways that science cannot register. Erikson notes that after such experiences, the vision of the world, of science and technology, nature and providence, changes.

> Post disaster, people often come to feel estranged from the rest of society and lose confidence in the structures of government...voices like those deserve to be listened to carefully.
>
> Erikson (1994)

Erikson's work is important to disaster studies as it adds clarity and humanity to the commonly found 'numbers game' concerning the 'magnitude of financial loss and number of deaths involved' (López-Ibor, 2005). Erikson classifies an event as disaster if it had the property of bringing about traumatic reactions. In his collection of accounts of very different forms of disasters, Erikson argues that a *new species of trouble* has emerged which blurs the line between the acute and the chronic, the natural and human induced. Examples include a valley in West Virginia known as Buffalo Creek, overwhelmed by a flood triggered by a dam collapsing; an Ojibwa Indian reserve in Canada called Grassy Narrows, devastated by methyl mercury contamination of the waters upon which they relied for their livelihoods and cultural identity; Immokalee, a small town in Florida where three hundred migrant farm workers were robbed of their savings; a group of houses in Colorado known as East Swallow, blighted by toxic fumes from pools of petrol that had gathered in the ground below; and the communities around Three Mile Island, site of the most serious nuclear radiation leak ever to occur in the USA. Erikson writes that such disasters, 'being of human manufacture, are at least in principle preventable, so there is always a story to be told about them, always a moral to be drawn from them, always a share of blame to be assigned.'

Erikson suggests however, that a *new species of trouble* often involves silent insidious toxins which 'contaminate rather than merely damage; they pollute, befoul, and taint rather than just create wreckage; they penetrate human tissue indirectly rather than wound the surfaces by assaults of a more straightforward kind. And the evidence is growing that they scare human beings in new and special ways, that they elicit an uncanny fear in us.' As Erikson notes in the preface to this book, the new species of trouble also destroys sense, and 'violates all the rules of plot' frequently associated with disasters. There is no 'beginning, middle and end.' The survivors at Buffalo Creek were overcome with a loss of sense, but the effect of silent destruction is even worse.

In a sense the disasters we face in the 21st century are not necessarily new: epidemics such as bird flu, extreme weather events such as hurricanes or tsunamis, malfunctioning technical systems as in train crashes, pollution from industrial processes, economic changes leading to unemployment, loss of biodiversity, or terrorist attacks from organizations inside or outside the country. What we hope is changing is the recognition that in a disaster, people want and need to be listened to, they need to be supported and appreciated. As Erikson indicates, when this assistance fails to materialise or is felt to be inappropriate or unhelpful, this can dramatically affect the 'way we see the world' a loss of sense in the orderliness of both nature and society. This frequently leads to an erosion of trust in governance and civil society. Following the 2005 floods, the accounts of the residents of New Orleans in the face of a seemingly incompetent and uncaring federal government (embodied in the woeful response of the Federal Emergency Management Agency (FEMA)) demonstrated this in a deeply moving and poignant manner (Lee, 2006).

In their excellent critique of Erikson, Law & Singleton (2004) note that disasters 'generate epistemological confusion and ontological uncertainty. Who am I? Why has order broken down? And then a kind of loss of morale, a kind of anomie that extends not only into the rules of the social but also into the natural order. There is demoralisation, and the destruction of communality.'

Ethnographies of disaster

So along with Blaikie *et al.* and Erikson, we agree that the people who can best describe the meaning and consequences of disasters are those who experienced them directly. In this way, the ethnographic approach may literally involve those people writing the disaster story. It then follows that such situated knowledge, based on localized, unfolding experience, is essential to the formation of effective and sustainable recovery policies and initiatives. But ironically survivors and those around them seldom get the opportunity to tell their stories. Either they are caught up in the process of survival having lost everything, or they may be already poor and lacking access or resources to communicate their story. They may be traumatized and unable to articulate, or believe that they are unable to articulate, their experience. Often they may be 'pathologised' and subjected to specific interventions such as counselling. They are viewed as victims rather than informants or sources of knowledge. Yet the telling of their experiences is vital to those outside the disaster, to recreate our conception of what it means to be human, to identify, for example, with what it's like to lose everything, or nearly everything. Erikson has spoken and written about those *inside* and *outside* disasters and the gulf in understanding which opens up between them. Survivors can easily become the *Other*,[6] a chaotic and needy reminder of loss and failure, or to symbolise dependence, frightening and dangerous in their raw desperation.

So how can people living through disasters tell their stories? Apart from the often reactive and transitory efforts of journalists, the means to do so has probably been taken away either by the disaster itself or responses to it. An exception here is the remarkable film by Spike Lee, *When the Levees Broke*, about the New Orleans floods which followed in the wake of Hurricane Katrina. Lee realized early in the project the power and articulacy of survivors' stories. The montage style film, which contains more than fours hours of testimony, tells the story for itself without recourse to narration or commentary, the filmmaker simply taking the role of witness. How then do researchers, and social scientists in particular, find ways to implement a commitment to helping survivors find and use their voice? As Marcus has discussed (1986), this is a particular challenge for the practice of ethnography as it moves away from its early colonialist beginnings towards a more committed and interventionist approach. How can ethnography itself both give account, and be accountable?

In this book we elaborate our own approach to these questions, through our work on a large scale 'lay ethnography'. It is for others to judge how far it has enabled the stories of those affected, the 'insiders' to reach the 'outside', any failure in this respect is entirely our own. In the 2001 Foot & Mouth Disease epidemic while there was no large-scale loss of human life, disaster in the sense of the community being overwhelmed was everywhere, and in some rural areas of the UK the normal patterns of life vanished for almost one year. The accounts we collected look back over this year and then record what everyday life was like in the subsequent 18 months.

Our approach has also been influenced by the Mass Observation movement which started in the 1930s. Pioneered by Harisson, Jennings and

Madge, their collection of diaries and reports from hundreds of ordinary citizens aimed to construct an 'anthropology of ourselves'. [7] In a similar way, rather than gather the views of 'key informants' or FMD 'experts' we wanted to hear everyday accounts from as wide a range of rural people as possible. But at the same time it was important that those stories could be given to us in as great a depth as those people, the 'lay ethnographers' wanted. We were clear that it was *the respondents* who decided what was important, it was *they* who decided how much to say and when to say it. So we decided to ask no questions about the FMD epidemic as such, but to let diary accounts emerge from current daily activities. It is the everyday stories from disasters which tell us how people coped and carried on, and which we would argue, becomes so moving. For example, a member of a slaughter team recalls the remarkable kindness of a farmer whose livestock was about to be culled:

> *We had an old man at Aspatria. He was absolutely broken hearted when we phoned him and there was 3, I think it was 3 or 4 wagons went* [to collect the slaughtered stock] *and he said 'make sure they come half an hour earlier and I'll get their dinners before they load the sheep' and he fed them, and when the sheep were loaded he took them into the house and gave them a cup of tea before they went. Because they were the same way, they knew it was a lifetime's work.*

We acknowledge that first in collecting and then in ordering and presenting these accounts we were 'co-ethnographers' charged with particular responsibilities. The accounts we give in this book, taken together with the public archiving of the material we have undertaken[8] are then, a collective re-telling of the disaster.

2
Global to Local: The Case of Foot and Mouth Disease in Cumbria

Introduction

> Cumbria in Northwest England was at the epicentre of the epidemic. It suffered 893 outbreaks[9] and was the second longest affected area. The County which is rich in natural heritage and scenically beautiful, has livestock farming, tourism and outdoor recreation among its economic mainstays. The effect of FMD was therefore devastating.
> (Cumbria Foot and Mouth Disease Inquiry Report 2002:9)

> BBC Radio Cumbria broadcast from the heart of the crisis. We were relied on for fast and accurate reporting of the facts, our ability to understand and capture the mood of the community, and our regular scrutiny of key decision makers. [...] We introduced a five minute foot and mouth bulletin that ran seven times a day at the peak of the outbreak.
> (Graham, 2001: 5)

The extract and quotation above convey something of the nature, scale and devastation of the 2001 FMD disaster in Cumbria and how it was experienced by Cumbrian rural communities, households and individuals. How did individuals and groups endure daily disruption and what was for many profound loss, and for many more, the fear of loss? For most there was an altered everyday reality that has left a lasting memory. Drawing on a wide range of sources, we chart both a collective story of what happened in Cumbria and individual stories, personal recollections from local people who worked closely with us. The collective and individual reveal indexical accounts of FMD (references to concrete events in time and place) and rich, diverse, sometimes contradictory narratives or stories that nevertheless became coherent when brought together. This storytelling, a co-ethnography of the 2001 FMD disaster captures both personally meaningful accounts of trauma and recovery and localized cultural context of experience. First, however, we start by briefly considering the history of FMD in the UK and some of the 'numbers' related to the 2001 epidemic.

FMD's social history and organisational context

Epizootics, or epidemic outbreaks of disease in an animal population, have for centuries caused devastation in farming communities across the world. There are a number of epizootic diseases which are of particular global importance, these include Avian Influenza, Epizootic Haemorrhagic Disease Virus, Rift Valley Fever, Rinderpest, and FMD. According to Ward *et al.* (2007) FMD is a highly contagious disease of cloven-hoof animals (among farm stock, cattle, sheep, pigs, goats and deer are susceptible). It is caused by a virus in the family Picornaviridae and genus Aphthovirus (the same family as the common cold virus, Rhinovirus). The virus, which has an incubation period of between 2–14 days, can infect in small doses, with a rapid rate of disease progression (Health and Safety Executive, 2007). Symptoms typically commence with fever, followed by the development of vesicles (blisters) mainly in the mouth and on the feet. The virus can be present to a high level in all secretions and excretions of acutely infected animals, including in their breath. It is probably more infectious than any other disease affecting animals (as few as ten infectious virus particles are enough to infect a cow) and spreads rapidly if uncontrolled. The most common route for transmission is through direct contact between infected and susceptible animals.

In 2001 FMD in the UK was not a new phenomenon. Reynolds & Tansey (2003) note that the first recognized case of FMD in Britain was in 1839 and statistical records show 27,254 cases during 1870.[10] Between 1922 and 1967 there were four major FMD epidemics with ensuing government reports in 1922, 1924, 1954 and 1967/8. The Gowers Committee of 1954 (a Committee of Inquiry Report to the then Ministry of Agriculture) reported that there had been no single year between 1929 and 1953 in which the UK was completely free of FMD.

The outbreaks of FMD during the late 19th century gave early indicators of the importance of economic markets and animal movements for the spread of FMD. This may be best illustrated by an early example of globalization; the case of large scale mutton and beef imports from Argentina during the last decade of the 19th century. The development of refrigerated sea transport and the advances in technology had shortened the journey times across the Atlantic meaning that both chilled meat and live fatstock imports became a thriving and profitable business with British citizens investing in Argentine railways, stockyards and processing infrastructure. However over the course of 40 years FMD had become endemic in Argentina and by 1900 had spread to the Buenos Aires area, the centre of the export trade. Britain banned live imports and Argentina responded by introducing livestock controls hoping that this would assuage British concerns. This was the beginning of a series of embargoes and restrictions affecting Britain, Europe and South America. Suspicion began to fall on chilled meat as a possible vector for FMD (when it was fed to pigs in unboiled swill). For the next 30 years the science, control measures, and trade bans continued to develop amid claim and counter claim as to the exact route and cause of each outbreak. During the Second World War years, Ministry of Agriculture, Fisheries & Food (MAFF)[11] officials admitted for the first time that South American meat was possibly to blame for many British outbreaks, information that had hitherto been censored and denied.

However, as Abigail Woods notes in her comprehensive history of FMD, 'A Manufactured Plague', the Argentines, for whom the disease was endemic, were bewildered by the stringent trade regulations brought in by Britain in the late 1920s. They saw FMD as a mild disease which caused temporary loss of condition but from which animals recovered and which could be controlled by isolation. Clinically the disease did not cause the severe symptoms in fattening stock that the British reported afflicted their dairy and beef breeding cattle. Woods (2002:61) reports that after reading British descriptions of FMD, many Argentines simply did not believe that it was the same as the mild ailment they called *'aftosa'*. '...some authors have argued that the Argentine failure to recognise FMD as a terrible plague was due to ignorance, this is unfair. While the Argentine understanding of FMD was very different from that which prevailed in Britain it was certainly not irrational. One of the main reasons why British farmers dreaded FMD was because of the hardships inflicted by measures used to control the disease. In Argentina there were no such controls and therefore the disease had very different social and economic implications. It also presented a far milder clinical picture. FMD symptoms were not particularly severe in fattening stock when compared to breeding and dairy stock and frequent bouts of infection would have left

Argentine cattle with a high degree of immunity. Furthermore, most Argentine cattle were reared extensively where they were handled infrequently and their condition monitored only occasionally. Living in a country where FMD was endemic and where infection spread mostly via the movements of infected stock, many Argentines simply did not understand the British fixation with disease importation.'

Writing about the UK 2001 outbreak in the Guardian newspaper, Woods (2001)[12] describes a relatively mild infection where the 'vast majority' of infected livestock 'suffer only lameness and loss of appetite, recovering in two or three weeks'. This would seem to be at odds with accounts of the infection encountered by farmers, vets and slaughtermen working on the front line in Cumbria in 2001. For example, a farmer reports a livestock cull at a neighbouring farm:

> *I mean you hear on the telly like that some of these experts saying 'Oh there's no need to slaughter them they would get over it' well there's no way these cows would get over it like. I mean their tits was dropping off and bits of their tongues was coming off they said were absolutely stinking when they shot them. Reeking.*

Similarly, a slaughterman describes what he first saw when he arrived at a farm:

> *The farmers wife she just wanted us to kill them because they were that bad. There was froth from the cows' mouths 6 to 8 inches on top of water troughs. They couldn't drink, they were that bad...they were absolutely wrecked. There was no reprieve for them animals, them cattle... their tongues was just like pieces of liver in no time at all. Just blistered and all...They would stand there with their tongues out, big blisters on them, blisters on their lips, and er round, you know where the hoof joins the fleshy part of the foot , all round there. Oh they were in absolute agony were them cows, absolute agony [at one case] they'd had it a while, they'd literally start rotting. You could smell them...some of these cattle, the teats had come off them...literally just rotted like. Terrible terrible disease.*

Farmers reported similar distressing symptoms in relation to the Summer 2007 UK outbreak of FMD. An interview with farmer John Gunner in the Times newspaper (8th August 2007), who had 102 of his cattle culled, quotes him as saying that he 'noticed that something was not right at lunchtime, one cow was limping and another dribbling at the mouth. By this point my old bull wasn't very well...in the short time I was there he collapsed in front of my eyes.' He said he was shocked at the speed of the transformation in the herd, 'they were all going skinny in front of our eyes. They were stumbling and falling over and drooling...it was just so fast.' Mr Gunner said that the DEFRA vet asked him if he would be prepared to cull his herd before FMD had been confirmed and he told them, 'you have got to do your job...it was a foregone conclusion, it would have been unfair to keep them alive.'

A Cumbrian vet with 20 years experience in large-animal practice, who worked on the 2001 epidemic, told us that: 'Within 24 hours of getting to Carlisle we were called to a farm late at night, two cattle clearly had FMD. Their suffering was so intense I couldn't wait to put them down. We looked through the rest of the herd and about 10% were showing signs of the disease. The following morning when the slaughter team arrived it was astonishing how ill most of the herd were in such a short time, they had high fevers and were in pain. On other cases I saw cattle that had sloughed off the entire lining of their tongues and were in terrible pain...the cattle would have been put down on welfare grounds even if they had not had FMD. I think the trauma suffered by some vets was due to the sight of animals in that state as well as the human suffering... although the methods of control of FMD may be seen as a reaction to possible economic loss, that view glosses over the severe animal suffering caused by the disease.'

This comment illustrates how economic interests attracted more attention during the 2001 disaster than the animal suffering encountered on the ground. Of course, as our own work shows, economic considerations certainly took precedence over the *human* suffering. This is, however, not to disregard the economic impacts of FMD, which can be long-lasting and substantial. For example, in a submission to the EU regarding the 2000 outbreak of FMD in Uruguay (whose economy relies largely on agriculture and agricultural exports), Paul Sutmoller (2002) notes that 'The temporary interruption of the export markets and the pronounced decrease of livestock prices proved very costly for Uruguay in terms of economic losses by the livestock production sector and the national trade. In particular, the financial losses of meat and dairy producers have had great negative impact on the national economy. In addition, the paralysis of the whole of the livestock sector, including the packing plants, for several months also affected many workers of livestock dependent activities. It is estimated that the total losses as a result of the closing of 70 external markets for products of animal origin exceed the total of more than 200 million US$. In addition, 380 containers with meat in transatlantic transit were returned for security reasons. This loss in combination with the losses related to the closed packing plants amounted to a total of $30 million US Dollars.'

There also exists a tension in countries which use vaccination as a control method but are then bound by the longer quarantine no-export periods laid down in the International Animal Health Code (see Anderson Inquiry, 2002:122–123). While Britain was debating and ultimately rejecting the use of vaccination as a control method in 2001, a pandemic of FMD in South America was being tackled by means of vaccination, isolation and moderate culling. In the early 1990s the EU harmonized its approach to vaccination and the routine use of FMD vaccine was eventually prohibited after a phasing out period. This duly influenced policy in South America, and led to a move away from routine vaccination towards 'FMD-free status.' Uruguay, Paraguay, large parts of Brazil, Argentina and Chile (which had been FMD free since 1988) were recognised by the *Office International des Epizooties* (OIE) as FMD-free. However as Sutmoller (2002) argued in his evidence to the EU subcommittee on FMD, this push toward disease-free status left these countries more vulnerable to any future outbreaks as animals lost their natural resistance, veterinary surveillance became less close and organised and there was insufficient planning for the transition to vaccination-free regulations. He goes on to describe how major outbreaks affecting Brazil, Argentina and Uruguay were tackled using vaccination and makes a case for the judicious use of vaccination as a control method suggesting that: 'The transition from a "FMD free with vaccination" status to a "FMD free without vaccination status" requires a different mind set of all stakeholders and the preparation of flexible contingency plans.'

With links back to the late 19th century and early 20th century outbreaks of FMD there were claims during the 2001 UK epidemic that the virus had reached the UK as a result of South American beef imports. However, as Gallagher *et al.* (2002) report 'The Pan-Asia strain of the FMD virus (which spread through UK in 2001) first appeared in 1990 in India. It had spread through middle east as far as Greece by 1996 and through Asia causing outbreaks in the far east during 1999 and 2000. This particular strain has never been identified in South America.'

In September 2000 the European Commission for the control of Foot & Mouth Disease convened an Expert Elicitation Workshop in order to answer key questions relating to the importation of the disease into

European countries from countries outside Europe (Gallagher *et al.*, 2002). The working group also considered where in Europe a primary outbreak of FMD was most likely to occur and the number of outbreaks expected to occur within European countries in the next five years. The Balkans group of countries was considered to be the most probable group within Europe to have a primary outbreak of FMD and also most likely to have the highest number of primary outbreaks. Turkey was considered to be the country outside Europe which was most likely to be the source of an outbreak within Europe as a whole, and the illegal importation of livestock was considered to be the most probable route of introduction of FMD into Europe. Results specific to the Islands group of countries, which included the UK and Ireland, suggested that this group was expected to have a mean of one primary outbreak of FMD in the five years from September 2000, and that the importation of foodstuffs by people entering those countries from Turkey was the most likely source of an outbreak (Anderson Inquiry, 2002).

Prior to this, on 18th July 2000 Jim Scudamore, then Chief Veterinary Officer had written to his deputies to say that, 'Having been at Pirbright for a day last week and seen the various maps of the deteriorating FMD situation in the Middle and Far East there appears to be an increasing risk of incursions from exotic viruses'. On the same day the Assistant Chief Veterinary Officer in Wales had written to the CVO to express his concerns about the lack of progress in implementing the recommendations of the Drummond Report.[13] Presciently he had stressed in particular the recommendations for staff training and the slaughter and disposal of stock. While these concerns were exchanged and discussed within the veterinary profession there was no action taken outside the State Veterinary Service to address them (Anderson Inquiry, 2002).

In 1991 the EU published 'Recommendations or Guidelines for Contingency Plans against Foot and Mouth Disease DGV1/1324/9' which advised that each country should have plans in place for the immediate deployment of enough trained and qualified staff to deal with up to 10 cases simultaneously and to monitor a 3 km protection zone around each. The 2002 Anderson Inquiry report questions the Government's assertion in its memorandum to the Inquiry that 'comprehensive contingency plans were in place.' Anderson states that these contingency plans were 'limited in scope, out of date in some respects and not integrated into a national programme of rehearsal and testing.' In addition, the report records that some stakeholders were unaware of the existence of the plan and one described it as the 'best kept national secret'.

Furthermore, it would appear that the existing contingency plans were both flawed and unimaginative when it came to the possible scale of a new outbreak. The EU guidelines were looking at a worst-case scenario of ten cases. In fact by the time the first 2001 outbreak in Britain had been formally diagnosed the disease was already incubating in more than 50 other locations between Devon and Galloway. The concerns of the Drummond Report and the Chief Veterinary Officer had not led to any improvement in the planning and preparation for a future outbreak of exotic disease. Perhaps it was simply that Britain had for so long been free of FMD that the scale and ferocity of the 2001 outbreak was inconceivable.

In contrast, however, the Netherlands had had recent experience of a catastrophic outbreak of Classical Swine Fever in 1997 and as a result had a strict emergency procedure in place: they imposed a transport ban on susceptible animals imported from UK on 21 February 2001 (the morning after the first confirmed diagnosis of FMD was made at Cheale Meats, Essex). On the same day the Netherlands banned sheep movements to markets, collection centres, auctions and exhibitions and on 22nd February they extended this ban to all susceptible animals. Meanwhile in the UK, auction marts were not closed until 17.00 the following day, Friday 23rd February, meaning that in the UK three full days of trading had been allowed to continue following the confirmed diagnosis, by which time the disease had been seeded the length of the country.

In autumn 2007 there was a further outbreak of FMD in Surrey, South East England. Most commentators seem to agree that this episode was managed more effectively than 2001. The National Farmers Union (NFU) President Peter Kendall, who gave evidence to the 2007 Anderson Inquiry, noted that although this was an outbreak that should never have happened, it had at least been handled much more effectively than the 2001 catastrophe.[14] 'That this outbreak should never have happened' is a reference to the source of the 2007 outbreak, which was almost certainly the nearby laboratory and manufacturing facilities at Pirbright. The irony is that the two main organizations based here are both heavily involved in FMD research and prevention. The Institute of Animal Health (IAH, commonly referred to as the Pirbright Institute) is the only United Kingdom facility licenced by Government to hold and work with live foot and mouth disease virus and acts as the National and World FMD Reference Centre; a role which includes international monitoring of FMD outbreaks. Merial Ltd is a global animal health pharmaceuticals company, licenced to produce FMD vaccine. The Pirbright

Institute has a long history of involvement with FMD. Writing in 2003, for instance, Reynolds & Tansey (2003:viii) note that 'with regard to FMD vaccines in Britain, it is curious that although State sponsored research (at the Pirbright Institute) followed the serious FMD outbreaks in 1922 and 1924, the resultant effective vaccines have never been used in outbreaks in Britain.[15]

That the 2007 outbreak should be traced back to poor biosecurity at the Pirbright laboratories is indeed surprising. The Health and Safety Executive (HSE) report into the outbreak (2007:3) states that: 'such was the condition in which we found the site drainage system that we conclude that the requirements for Containment Level 4 were not met, thus constituting a breach of biosecurity for the Pirbright site as a whole. Our conclusion is supported by the evidence we found of long-term damage and leakage, including cracked pipes, unsealed manholes and tree root ingress.' The report also states that it was: 'likely that waste water containing the live virus strain, having entered the drainage pipework, then leaked out and contaminated the surrounding soil...excessive rainfall may have exacerbated the potential release from the drain.'

The disease is likely to have been spread due to 'inadequately controlled human and vehicle movements at Pirbright', and in particular construction vehicles, 'which were likely to have had unrestricted access to the site. In our opinion, the construction activities on site above the vicinity of the intermediate section of the effluent drainage system are likely to have caused disturbance and movement of soil in a way that contaminated some of the vehicles with live FMD virus. We conclude it likely that those vehicles, having driven over this part of the site, carried off out of the site materials containing the live virus in the form of mud on tyres and vehicle underbodies' (HSE, 2007:4). What is perhaps most shocking in this evidence, particularly following the stringent biosecurity measures imposed during and after the 2001 outbreak, is that the Government laboratory responsible for making and researching FMD apparently did not insist on washing vehicles on and off site and allowed the virus to enter the drainage system. One might reasonably assume that there is a more secure way of disposing of the FMD virus than sending it down the plughole.

Farming before FMD

Devastating though the 2001 FMD epidemic was, it would be disingenuous to argue that agriculture was problem free before FMD.

Underpinning the events of 2001 and during the last decade, the UK agricultural sector suffered significant problems (Franks, 2002; Lowe *et al.*, 2001; Report of the Policy Commission on the Future of Farming and Food, 2002; The Royal Society, 2002; MAFF, 1999). As the Royal Society Inquiry into infectious diseases in livestock states (2002:9), from the mid-1990s, 'much of the profitability has drained from the industry'. Contributory factors have been the strong pound, an excessive supply of sheep and the Bovine Spongiform Encephalopathy (BSE) beef market crisis. Further public health scares such as E. coli 0157 have undermined public confidence in large-scale food production. Rural economies were thus under pressure before the onset of FMD so that by the time of the epidemic, farm incomes were 'on the floor' (Report of the Policy Commission on the Future of Farming and Food, 2002:13) and the industry was already coping with problems of poor public image, rising average age of farmers and profound uncertainty about the future.

Within Cumbria, the 2001 FMD epidemic massively damaged livestock farming that was already struggling to survive.[16] This was particularly so in the more remote, upland parts of the county characterized by the lower income hill farming sector.[17] A damaged farming economy negatively impacts on related economies, and tourism – a key sector of the Cumbria economy, is often inextricably linked with farming, particularly in areas of outstanding beauty such as the Lake District National Park. Many farms also offer bed and breakfast accommodation, as well as other activities such as pony trekking and farm walks. In addition, there are numerous footpaths and bridleways that run through farms and grazed commons (Bennett *et al.* 2002, Lowe *et al.* 2001). The virtual closure of the countryside for almost a year brought much of this tourist industry to a standstill.[18] Distressing media portrayals of burning carcasses compounded the problem. The reduction in visitor numbers to rural areas, meant hardship and anxiety for those running tourist/visitor oriented businesses. Many village shops and pubs, upon which rural communities rely all year round, are themselves reliant on additional income from visitor trade for their survival, demonstrating the delicate interrelationship between these particular constituencies. Here a rural vicar recounts:

> *One of the significant things to emerge from FMD as far as my rural parishes were concerned was the almost total disappearance of visitors. My most visited church (…) has just ¾ page of entries in the visitors book between March and September 2001, as opposed to 35 pages the year*

before! The knock-on effect for the pubs, B & Bs and the local shop were huge, indeed (one of the two village pubs) is now temporarily closed...the landlord has gone into voluntary liquidation and the Post Office/shop is struggling, takings were down by over 50% last year! A stark indication of how significant tourism is to our small patch.

FMD numbers

It is at times difficult to comprehend the sheer scale of the 2001 FMD epidemic. As Bennett et al. (2002) indicates, this was the world's worst recorded epidemic of the disease and the most serious animal epidemic in the UK in modern times. This was the first outbreak in the UK for 34 years, an event described as the most serious ever to occur in a previously FMD-free country (Cumbria Inquiry, 2002) and 'a traumatic and devastating experience for all those who were affected by it. It was a national crisis' (Anderson Inquiry, 2002). Here are some of the key points.

FMD was suspected at an abattoir in Essex on 19 February 2001 and confirmed the following day. The epidemic lasted for 32 weeks, the last case being confirmed on 30 September 2001 on a farm near Appleby in Cumbria. The impact of the epidemic was immense: much greater than that of the last major outbreak of the disease in the UK (which occurred in 1967–68). In mid-April 2001, at the height of the crisis, more than 10,000 vets (many contracted from all over the world), soldiers, field and support staff, assisted by thousands more working for contractors, were engaged in fighting the disease. The scale of animal slaughter in Cumbria was unprecedented at the peak of the disaster. One site alone, a disused airfield in the village of Great Orton, near Carlisle in Cumbria, was dealing with the daily slaughter and disposal of up to 18,000 animals (this was the largest disposal site used during 2001).[19]

On 14th January 2002, after a cumulative total of 2030 cases,[20] the culling of between 6.5 million and 10 million animals,[21] and economic loss which the Department of Environment, Food and Rural Affairs (DEFRA) estimates to be £9 billion,[22] the last county (Northumberland) was declared FMD free (Anderson Inquiry, 2002, Picado et al., 2007, Campbell & Lee, 2003). Overall in Cumbria, outbreaks endured from 28th February until 30th September, with individual farms continuing under restrictions into the spring of 2002. In North Cumbria 893 farms had confirmed infected cases (44% of the UK total), with a further 1,934 having complete or partial culls of livestock (National Audit

24 *Animal Disease and Human Trauma*

Office, 2002). In the north of Cumbria, 70% of farms experienced livestock culls. Figure 2.1 indicates the location of Cumbria within the UK. Figure 2.2 indicates the spread of cases across Cumbria and Figure 2.3 shows the distribution of cases in the UK.

Tourism suffered the largest financial impact from the outbreak, with lost revenues of between £4.5 billion and £5.4 billion. Visitors to Britain and the countryside deterred by media images of mass pyres and the initial blanket closure of footpaths by local authorities (no entry signs, such as shown in Figure 2.4, become a common sight in

Figure 2.1   The location of Cumbria (shaded area) within the UK

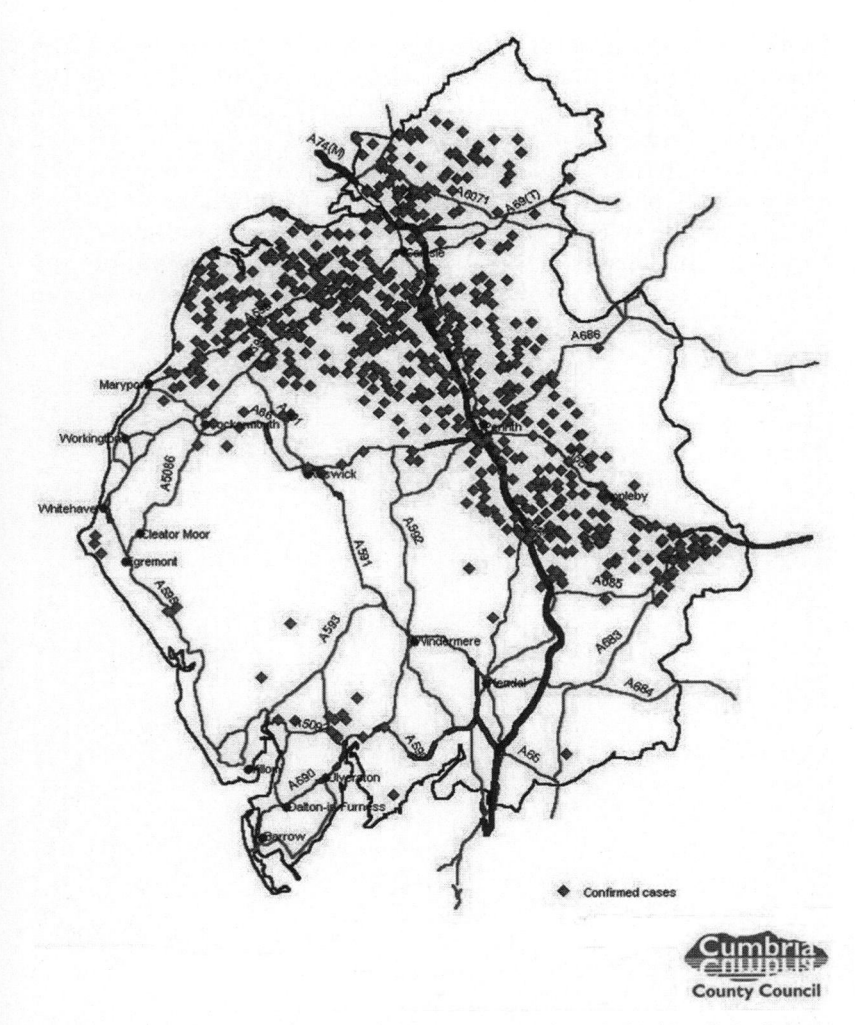

Figure 2.2 The distribution of FMD cases in Cumbria 2001 (Cumbria County Council, 2001)

Cumbria). On 22 January 2002 the United Kingdom was re-instated on the OIE1-list of countries free of foot and mouth disease, and on 5 February 2002 the European Commission lifted remaining meat and animal export restrictions.

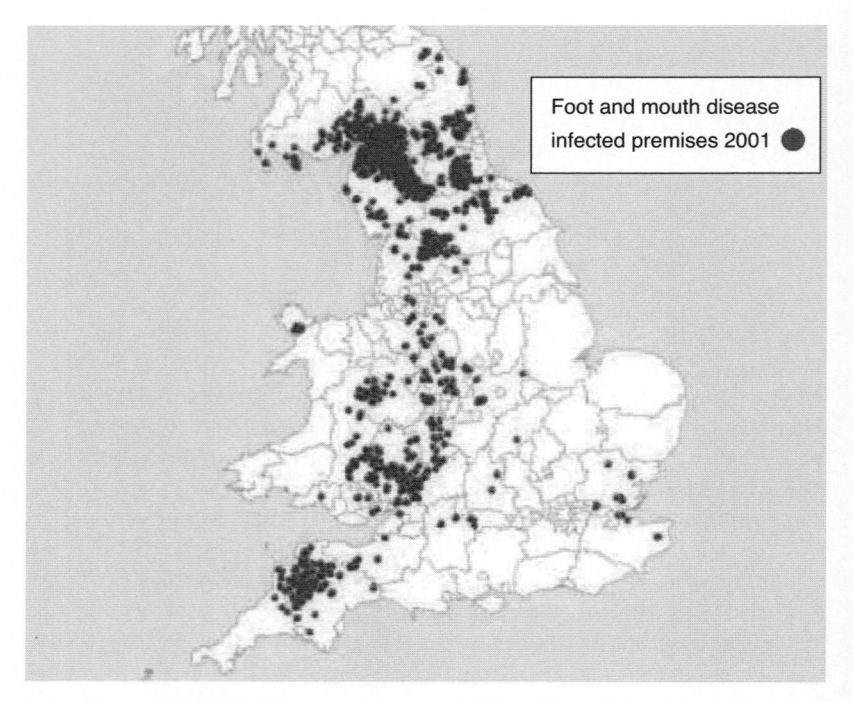

Figure 2.3 Distribution of FMD cases in the UK 2001 (adapted from Cumbria Inquiry, 2002)

Figure 2.4 'No entry' Cumbria 2001

A disaster unfolding

> Today Cumbria is still seared upon my soul. Air crashes, road accidents, riots, the odd murder, nothing in 25 years of reporting domestic news had ever prepared me for the scale of human trauma I experienced that mad March weekend last year. It was unutterably brutal.
>
> (Jon Snow, Forward, Cumbria's Story, The Cumberland News 2002:3)

The journalist Jon Snow's 'mad March weekend' recollects his 2001 visit to Cumbria to compile a special FMD report for Channel Four News. This was some five weeks after the confirmed FMD case at Heddon-on-the-Wall, Northumberland, 30 miles from the Cumbrian border. Within those five weeks, an outbreak at a sheep dealer's farm in Devon had been linked to the Heddon case via Longtown auction mart in Cumbria, one of the biggest marts in the country. Moreover:

> Longtown [mart] was at the epicentre of a maze of 25,000 animal movements across the UK between February 13th and 16th. By Thursday March 1 Cumbria's first case was confirmed at Smalmstown near Longtown.
>
> (Commentary, Cumbria's Story, The Cumberland News 2002: 8)

Nationwide livestock movement restrictions followed,[23] and the UK Ministry of Agriculture Fisheries and Food (MAFF)[24] had a national 'stamping out' control strategy: the slaughter of susceptible animals from infected farms and from farms considered to have been exposed to infection (dangerous contacts) which were then buried locally or incinerated on pyres.[25] As we now know and as the above quote illustrates, the problem was that the disease had not been detected for some three weeks and there were problems tracing nationwide livestock movements. There is evidence from our study that though movements were all documented, MAFF/DEFRA were unprepared for the scale of the task of tracking down the stock. For example, a Longtown haulier notes:

> *They asked us for our movements from the day before Hexham till the day the transport stopped, and we just done a printout off the computer and they phoned back and said we only wanted a week or fortnights movements. We said well that is what it is. That is what we moved, we were*

> moving anywhere between 10 and 18,000 head of livestock a day. And they just, they wouldn't believe it. Because transport grew that much, the roads grew and vehicles got better, you can cover that much distance in a shorter space of time.

This perspective is echoed by George Studholme, a slaughter man working in Cumbria, who recounts that in February 2001, everyone watched the news with increased interest when it was announced that infected animals had passed through Longtown Auction, the largest sheep auction in the country and we knew that the 'shit had hit the fan'. Dealers from all over the country traded at Longtown and most farmers in Cumbria and South West Scotland used the auction (extract from personal account recorded in Graham, 2001: 21).[26] The Government's next step did little to restore confidence. On March 16th, Nick Brown, then Minister of Agriculture, announced to Parliament new measures[27] to control the spread of disease, which included the culling of all animals within 3 km of every foot and mouth outbreak. Whilst there was some support for culling sheep thought to be incubating the disease, there was stunned disbelief that beef and dairy cows, most of which remained housed at that time of the year, were also to be slaughtered.

> On Thursday March 16 at 1.30 pm Brown unveiled the next step. A huge cull of 'all animals' in 3 km zones around the infected farms was the way to get ahead of the epidemic.
> Commentary, Cumbria's Story (The Cumberland News, 2002:20)

As the Cumbria Inquiry (2002) notes, the Government's announcement of the 3 km cull was poorly handled and, as initially made, was in error. In some instances the letter informing individual farmers of the decision to slaughter their sheep failed to provide a satisfactory explanation for the action being proposed. There was confusion amongst farmers about whether the slaughter was voluntary or mandatory, and an understandable perception that the authorities were acting threateningly. Moreover, the quantitative scientific justification for the 3 km culls has never been fully set out and many vets in the field were strongly opposed to the idea. As one vet heavily involved with the inspection of livestock put it:

> Nick Brown stood up and he said he was going to slaughter everything in Cumbria that was within three kilometres. He meant it. He meant it.

Everything, cattle, sheep, pigs, everything within three kilometres. And there were dead bodies everywhere, there were twelve report cases that farmers had phoned in saying, we think we've got foot and mouth, we'd like a vet to see it, and there weren't vets to go and see them.

Confusion reigned as farming unions sought clarification of the meaning of 'all animals' from the Government and it was not until a local television interview that evening with Brown, that the situation was clarified. 'All animals' included sheep and pigs, not cattle.

It became clear fairly quickly that the policy of 'stamping out' was not working, and on the basis of some of the epidemiological computer models produced by the FMD Science Group, Professor David King, the Government's Chief Scientific Advisor, advised the Government on 23rd March to modify the disease control policies and to introduce what became known as the '24/48 hours contiguous culling policy'. This involved slaughtering stock on infected or suspected, infected farms within 24 hours and stock on adjoining farms within 48 hours.[28] We now know that one of the most important effects of this modeling was an overestimate of the significance of local spread, resulting in an unjustified high number of premises being pre-emptively culled in some areas (Picado et al., 2007). This process has been referred to elsewhere as 'carnage by computer' (Campbell & Lee, 2003).

On the ground, slaughtering on this scale caused a logistical nightmare in terms of culling and disposal of livestock. As early as week three from the first confirmed case in Cumbria, the local MP David McClean pleaded with the Government for more resources including support from the army, but the Minister of Agriculture, Nick Brown, and the Government's Chief Veterinary Officer, Jim Scudamore, insisted that it was 'under control'.[29]

Moreover despite the policy of contiguous culling,[30] outbreaks persisted and there were fears that this 'science driven' approach lacked common sense and alienated and marginalised local knowledge. Farms 'taken out' as part of a contiguous cull were seen to be plotted on a map by 'officials in London' without due attention to local situated knowledge. For example, Bickerstaff et al. (2005) draw on findings from focus group discussions in the south west of England, to consider local experience of the 2001 FMD disaster from an environmental risk perspective. During one group discussion, it was suggested that a neighbouring farm was 'taken out' because on a map, it seemed that it 'joined' an infected farm along a 50 yard stretch of a river. However the river lay between the farms and the cattle were housed indoors.

Also in a manner resembling the mismatch of knowledges which undermined central agency credibility in the management of the Chernobyl fallout in (Wynne 1997) local practices were not taken into account. For example, one farmer told us:

> They went around in helicopters looking for sheep because they thought that this other half million sheep was there, but this was like in May in lambing time. They were going on Figures in June from two years previous, which there would be that many more sheep, because all the ewes are on then and the lambs would still be running about, but try to tell them, they just wouldn't have it.

By the time of Jon Snow's visit in March, there were 227 confirmed FMD cases in Cumbria, just under a third of the then current UK cases (Cumbria Story 2002:30). Within the county, there had been dramatic increases from week to week: 2 confirmed cases by week two, 40 by week three, 90 by week four. On Monday March 19th Jon Snow's special FMD bulletin was broadcast from a Carlisle hotel roof top; by the Friday of that same week the then Prime Minister, Tony Blair was in Carlisle, 'promising unlimited resources and to cut through the red tape and rules' and writing for a local newspaper, pledging 'we will not walk away.' (Cumbria's Story, The Cumberland News, 2002:30 & 40). On 22nd March, Blair became more centrally involved in managing the effort to control the disease, and established the Government's crisis-management committee 'Cobra', based in the Cabinet Office, in a co-ordinating role (Lowe *et al.*, 2003). By the end of March, the army was brought in under the command of Brigadier Alex Birtwhistle. The following excerpts (from the Anderson Inquiry and Cumbria's Story respectively) discuss the deployment and initial actions of the army:

> On 20th February the Ministry of Defence (MOD) was warned by MAFF that it might be called on for help. In response to MAFF request for assistance, the MOD deployed 'a number' of military vets on 14th March. Over the weekend of 17th March, 2 army logistics experts went to MAFF headquarters to scope what contribution the forces could make. That weekend resulted in deployment of military personnel to Devon on the 19th March and Cumbria on 21st March.
>
> Birtwhistle came the following week and on 26th March the first sheep, from the backlog of unburied culled animals, were brought to Great Orton, he reckoned that there was a backlog of about 70,000 sheep some of which would go for rendering with the 15,000 cattle

which would have to be incinerated because of the risk of BSE. At the same time, the landfill site at Hespin Wood near Todhills was identified for carcass burial. At 4 pm, March 27th, the Contiguous Cull became compulsory[31] and there were 20 slaughtermen at Gt Orton to kill the live sheep going on the cull and the Kingstown abattoir at Carlisle was working round the clock killing 10,000 a day. They were taken from there to Great Orton, Flusco (near Penrith) and Hespin Wood.

For many people the deployment of the army was a watershed in the management of the disaster, as the following extracts (from a farmer, a haulier and a slaughterman respectively) indicate:

It was only when the army came in that people breathed a huge sigh of relief.
It was good when the army got involved. It wasn't a case of oh can we do that, it was just do it, think about it later. You know, this is what we need, get on with it.
I mean, once they brought the army in, all right they didn't stop the Foot & Mouth right away, but I mean they did start to get shot of the animals and what have you and clean up the farms and there was none of this lying a week before they were moved or anything like that, they were shot and moved the same day.

The Cumbria Inquiry (2002:30) also reports that the Army contingent 'became an integral part of the disease control operations...During the Inquiry we were told on numerous occasions that there had been significant improvements in the management of disease outbreaks and in local relationships as the epidemic progressed. A number of witnesses commented favourably on the improvements that had followed the involvement of Brigadier Birtwhistle and his men. Though there were also some criticisms of the Army (including both administrative and operational failings), the Cumbria Inquiry concludes that in 'Cumbria, if not elsewhere in Britain, the Army should have been brought in sooner.'

In response to the national government decision not to carry out a public inquiry into the handling of the disaster, regional inquiries were held by major local authorities, the county councils, in Northumberland, Devon and Cumbria. The Cumbria County Council's Foot and Mouth Disease Inquiry Report (Cumbria Inquiry, 2002) outlines three aspects of the epidemic which had devastating effects on the county:

- A nation wide ban on livestock movements and widespread restrictions on public access to the countryside crippled Cumbria, a county

rich in natural heritage with an economy dependent on livestock agriculture, outdoor recreation and tourism;
- The massive scale of the slaughter and disposal of livestock and other animals. As well as the 893 infected premises a further 1,934 suffered complete or partial animal culls, taken out as dangerous contacts or in the cull of contiguous premises. Some 45% of Cumbria's farm holdings suffered these culls and in the north of the county where the epidemic was most severe, this rose to 70%. Also all Cumbrian farms endured stringent disease control measures that caused major disruptions to livestock management and in some cases, led to animal suffering and huge economic losses;
- The Government's handling of the epidemic: 'there were problems in implementation of disease control, communication and other measures [that] led to an upsurge of public objection and to expressions of public concern, frustration and anger at the way the epidemic was being handled.'

Whilst others note that the Government seem to frame the FMD epidemic solely as an agricultural problem (see for example, Ward *et al.* 2004), the scale of culling and disposal, together with the logistical exercise needed to carry it out, could not fail to have an impact on the communities involved. Indeed, the increased isolation for farmers which was the inevitable consequence of the policy of slaughter, disposal and bio-security; the virtual collapse of tourism as a result of restrictions on rights of way, and government advice to stay away from the countryside and the huge numbers of non-farming people drawn into the crisis, turned a veterinary and economic crisis into a human disaster for large parts of rural Britain.

Researching the disaster

The FMD epidemic in Cumbria in 2001 was a disaster in the sense that its speed and magnitude outran the ability of the existing resources to contain and manage it. The disruption and suffering were felt far beyond FMD's core location on farms and disposal sites. It was for this reason that we situated our research in North Cumbria where the difficulties were felt most intensely, in greater numbers and for longest, and where the effects spread most widely into the non-farming population.

While for some it was merely inconvenient, for others it caused lasting trauma, severe financial hardship and suspension of community life. As we will examine in later chapters, all the effects of community disaster could be seen at different levels of severity and visibility, and in communities in every sense of the word: geographical, cultural, occupational and demographic.

In the early stages of FMD, most of the attention about human effects was taken up with debates about the possibility of zoonosis (any infectious disease that may be transmitted from animals, wild or domestic, to humans). Workers and farmers were closely exposed during slaughter and disposal, and there were moves to reassure the public that 'foot and mouth disease doesn't cause a problem in humans.'[32]

However, amid all the debates and arguments about zoonosis and the possible (mainly physical) health effects of the pyres on which carcasses were burned, the social and affective nature of what was happening remained largely invisible. How could this happen? After all, this is an event described by Anderson (2002) as 'one of the greatest social upheavals' since the Second World War. It was abundantly clear to those involved at the time that the duration and scale of the crisis was unprecedented. Indeed, some media outlets asked farmers to send in a regular column to give the outside world a glimpse of what it was like to be cut off. This kind of immediate, often raw testimony stood out from the careful and guarded statements from government ministers, veterinary experts and agencies handling the disaster. By viewing the disaster through the lens of local expertise and experience, we hoped to better understand how the health and social consequences of the disaster went largely unnoticed, certainly amongst the statutory services.

These following diary extracts were written in the 'recovery year' of 2002. In the first extract, written in March that year (13 months after the first outbreak) a respondent tells a story of looking back, triggered by a task in the present.

> *I decided to sort out all the paperwork and guidelines we had received from DEFRA over the past twelve months – it was quite a pile!! It was interesting looking back and very sad, it makes you realise just how deep the feelings went and although it sounds silly it's surprising how quickly those feelings can return over the smallest things.*

The second narrative again is rooted in the present activity of gardening, but the writer conveys how the meaning of the activity has changed for villagers, how it takes on a new deep and poignant significance. It also speaks about the changed perception of and relationship with everyday places, a theme we return to in our discussion of altered lifescapes later in this book.

> Just over a week ago, R and I were busy getting our garden ready for the Village in Bloom. It was good to see everyone making an effort to everything looking nice. Everything felt a bit more normal this year, whereas last year I think the Village in Bloom was the last thing on most people's minds! It made me feel more confident about the future! Perhaps everything or most things may return to normal eventually!

In all we collected more than 3,000 weekly diary entries from a citizen panel of local expertise (54 people living and working in rural Cumbria).[33] The use of both citizens' juries and citizen panels as a consultative mechanism is well known in health and multi-agency groups, and has been used in assessments of health needs, (e.g. the Burnley Citizens' Jury, Kashefi & Mort, 2000). This procedure has been used by others formulating citizens' juries in other contexts (Coote & Lenaghan, 1997). However, whilst highly deliberative, these have been criticized for their lack of 'follow-through' and opportunity for learning (Dowswell *et al.*, 1997); (Harrison & Mort, 1998). It was felt that in order to utilize this approach in a *research* context, a longitudinal approach would be more productive.

The citizen panel was developed with the help of a multi-agency steering group, based on a demographic and occupational sampling frame. The panel members were purposively sampled from six groupings identified by the steering group as affected in a range of ways in the disease outbreak (see Table 2.1). The panel members became 'ethnographers of their own lives' and some of these individuals have retained their diaries and interviews as a chronicle of the 18 months following the world's worst epidemic of FMD, to pass down to children and grandchildren. Each member of the citizen panel drew on an individual lifescape, often as a way of trying to make personal sense out of the chaos and trauma that signified living with the disaster.[34] So as one respondent positioned herself as belonging to a Cumbrian livestock farming family trying to bear the culling of healthy sheep: *there was not a thing we could do*, another respondent who had to close their caravan and fishing lake site, aligned themselves to the lifescape of running a small rural business: *it happened to so many of us, businesses like us just closed*.

Of course the researcher is looking for patterns and themes and the panel members themselves became interested in this gathering process,

Table 2.1 Occupational and residential groups included in the rural citizen panel, and their study data

		Study data		
Group No	Members	Group discussions	Interviews	Weekly diaries
1	Farmers, farm workers, and their families	2	9	601
2	Small businesses, to include tourism, arts and crafts, retail, and others	2	8	394
3	Related agricultural workers, to include livestock hauliers, agricultural contractors, and auction market staff	2	9	576
4	Front line clean-up workers and managers (from DEFRA or Environment Agency), slaughter teams (temporary, Seconded, and permanent staff)	2	7	380
5	Community, such as teachers, clergy, residents near disposal sites	2	9	533
6	Health professionals, such as general practitioners, community nurses, and vets	2	9	587
	Total	12	51	3071

this other kind of sense making. As Ferdman (1990:185) notes, story lines recurrent amongst a range of contributions can shed light on the 'inter-relationships between collective and individual experience and behaviour'. Garaway (1996) suggests that narratives which have common elements can make cross case comparisons, which in turn may strengthen the 'legitimacy of emergent analytical themes'.

In addition, for many members of the panel the process of writing a weekly diary was helpful in coming to terms with the consequences of FMD. People told us that diaries offered a therapeutic space for reflection and an opportunity to make sense of experiences (see Box 1.1).

The researcher as co-ethnographer

The role of researcher as co-ethnographer raises some awkward questions. If we want to privilege testimony and let the 'stories tell themselves',

> **Box 1.1 Diary writing reflections**
>
> *It has been a tool by which I feel I have been able to mark/measure our (me and my family) progress/recovery from FMD*
>
> *It has been a chance to reflect on what happened, to look to the future and to recognise the frustration and anger*
>
> *Yes, this process has been useful in a number of ways and has helped considerably in coming to terms with the effects of the epidemic and its effect both on personal and business level. It has helped me stop and look at the future, and what I want to be doing in the future, or maybe helped provoke my mid-life crisis!*
>
> *I wouldn't go as far, to say I have enjoyed writing the diary, but it has certainly been a welcome form of therapy.*
>
> *I hope this has helped with understanding what people went through and hopefully if it ever happens again people will be more understanding about the pressure that everyone has been under.*
>
> *Source*: Mort *et al.* (2004)

what exactly is our role? Put simply, we are actors who make testimonies travel in different ways, through research presentations, reports, journal papers and artifacts like the ethnoplot (Figure 2.5.). The ethnoplot has been likened to a 'fingerprint', helpful in understanding the complex relationship between the 'external world' and the testimonies of respondents.[35] In the example below, self-reported health and quality of life have been combined with diary entries to track changes in a farm workers life over a period of 18 months. This way of combining qualitative longitudinal information with quantitative measures within one visual field could, we believe, be used both in other longitudinal research. It has potential for future co-ethnographic inquires, including the management of chronic illness, for instance dermatology where interactions between conditions, emotional states and/or food might be tracked over time; diabetes where self-care/control of blood sugar is known to relate to daily living activities and food practices, and bereavement, where different phases such as shock, grief and anger, require different approaches in counselling or self care.

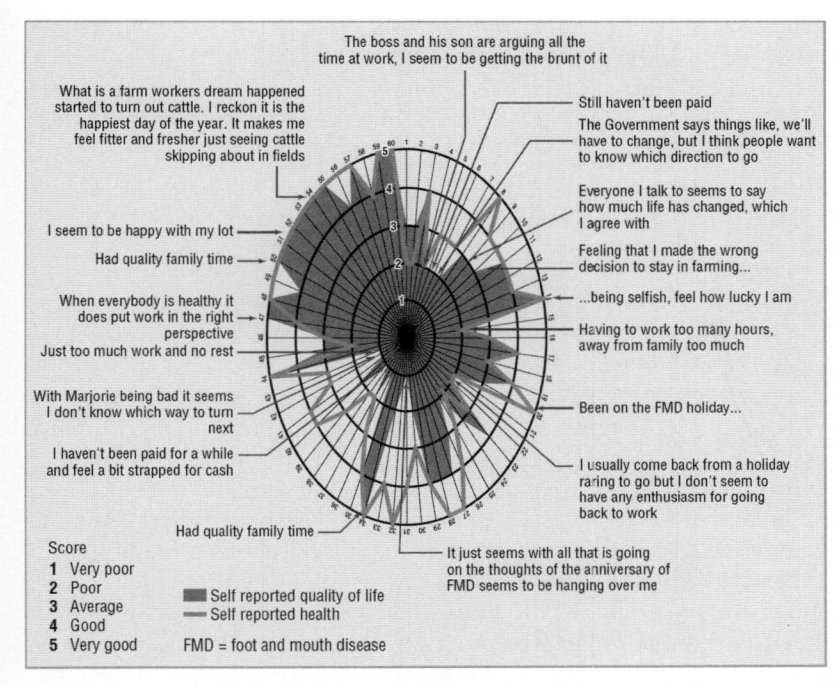

Figure 2.5 Ethnoplot of self-reported quality of life and health, together with diary data 'mapped' from December 2001 to May 2003 (Mort *et al.*, 2005)

The ethnoplot example above has been developed from the diaries, group discussions and interviews given by a farm worker who had reported severe distress and disruption during the epidemic. Diaries began in December 2001 (Point 1 on the *ethnoplot*), and early in 2002 he speaks of arguments at work (the farm) and about his wages not being paid, and uncertainty around the future of farming. These all correspond with times when he reports poor quality of life and changes in self reported health from poor to very good. Quality of life remains low when he says he feels he had made the wrong decision to stay in farming. This occurs again when he is reporting working very long hours and not seeing his family (wife and young sons). Next he reports going on 'the FMD holiday'. This corresponds with better health but is followed quickly by: *I usually come back from a holiday raring to go but I don't seem to have any enthusiasm for going back to work.* Quality of life is at its lowest (and health is fluctuating) at the anniversary of the cull on his farm of animal he had reared. *It just seems that with all that's going on the anniversary of FMD seems to be hanging over me.*

A brief episode of *quality family time* is linked with marked improved health and quality of life, followed by two very difficult months when lack of money, his wife's illness and very hard work, combine to give much lower scores. There then follows a gradual climb in health and life quality which is sustained and linked with wife's recovery, spending time with family and finishes with:

What is a farm worker's dream happened, started to run out cattle. I reckon it's the happiest day of the year. It makes me feel fitter and fresher just seeing the cattle skipping about the fields.

We can also see in this 'fingerprint', that there are points where quality of life and self-reported health do not overlap but are perceived or experienced differently. Quality of life at weeks 1–11 rates between 'poor' and 'average', yet self reported health climbs during this time to 'very good'. This divergence occurs again at weeks 30–33 and 40–44. Taken with the free text entries, the relational picture which can be shown in the *ethnoplot* offers, we believe, a more nuanced view of 'health' than approaches epitomized by standard questionnaires, which tend to assume that quality of life and self-reported health are convergent (Mallinson, 2001; 2002).

In practical terms, co-ethnography requires regular contact with respondents and a deep understanding of their personal situation. In addition to the usual group meetings and interviews, the diary approach mentioned earlier lends itself well to regular (usually monthly) contact between the co-ethnographers. Our monthly visits often took place in the respondent's home with other household members' present, or less frequently, in the workplace and a small payment was made to compensate for the time taken to complete the diaries. During these visits, conversations could be wide ranging, from discussing local and national FMD developments and initiatives; to everyday talk about families, paid work, past and future events, hopes and fears. Sometimes we would seek clarification from the respondent, of what had been written in diaries or possibly spoken of, during previous telephone calls (to arrange visits). In this way, regular visits place respondents in a position quite different to that of the everyday storyteller. Rather than telling stories that are 'of the moment' the research interview and diary approach seeks a biographical self-story in which diverse story fragments, events and anecdotes are brought together in an organized 'artefact'. If we are to privilege testimony in this way, we must then consider the nature of narrative in telling and retelling a story. From our experience and the work of others, we know that narratives rarely simply 'reveal' what someone thinks or feels, any 'truth' is a construction.[36] Biographical stories are always positioned, never complete.

The construction and portrayal of self comes about through the use of different narratives, depending on place, setting and audience (Richmond, 2002). Burr and Butt (2000) put it this way: 'we do not simply recall events as they happen; rather, we selectively recall, narrating a story of the past that makes sense to us'. Reissman (1993:218) also suggests that a sense of a beginning, middle and end, a linear and ordered telling of a tale is a storyteller's way of creating order out of a flow of experiences in order to make sense of actions and events. Ricoeur (1991:31) notes that we like to speak, write and recount in 'wholes' but story telling is always a pulling together of the 'parts of the discordant concordance.' Gubrium & Holstein (1998) elaborate on this 'pulling together of the parts'. They use the term 'narrative practice' to describe the dynamic processes of story composition, that which draws upon personal experience and available meanings, structures and linkages that comprise stories.

Thus a narrative offers a translation of a perception of events, presenting cause and effect in a selective manner, which makes sense to the author. Constructing and composing narrative always takes place within dynamic, shifting contexts. The longitudinal design of the FMD study, with participants weaving an ongoing narrative through 18 months of diary keeping, captures 'inconsistencies', 'contradictions', re-orderings and re-tellings that we suggest, may also be reflective of the chaos/farce that participants tell us they endured, during, and following the epidemic, and which they sought and still seek, to make sense of. These narratives may also reveal assumptions about 'shared structures' and common understanding of the cultural shorthand that describes these. For example, similar domestic arrangements assume similar narratives: *'you know what teenagers are like'*. Also assumptions about remaining outside of other structures, such as the relationship with one participant who described in great detail his working practices as a livestock haulier – narrative as instruction – *'so's you knows exactly what it's like for us, what the pressure's like dealing with all the paperwork.'* Moreover, visiting households meant that at times, others (e.g. partners, other family members) supplemented the material given by the participants.

A collective telling – a shared FMD narrative

We also needed to consider *how* a story is put together, the linguistic and cultural common elements that signify a shared narrative across a range of personal stories. Labov (1997) has argued that narratives have structures, formal properties such as an abstract, orientation, sequence of events, significance and meaning of the action and attitude of the narrator, resolution and coda (a concluding part). Whilst Labov's model may be criticized for omitting the relationship between teller and listener , it can provide a useful framework for understanding repeated story lines across a range of participants. With FMD, narratives revealed indexical recollections (references to concrete events in time and place), so that the stories themselves may be framed by local and cultural understandings of these events. They also have common elements such as memories of hearing about the first FMD outbreak (beginnings); the disruption of *normal* routines, which for many participants led to feelings of discord – everyday rhythms becoming disharmonious; in turn fostering a sense of separation, disjuncture–feelings of being disconnected and finally in some cases, distress arising from an overwhelming sense of loss, hopelessness and despair.

These common elements provide the framework for a collective telling of a shared FMD narrative. They capture not only the common plot but also individual perceptions of actions and outcomes. The narrative is a 'pulling together of the parts'. Moreover, the telling of the story, its context, highlighted common use of significant language, such as 'clean/dirty', 'inside/outside' that has specific meaning to the 2001 epidemic. Outside of this context, such language is not so emotionally charged. Whilst there is a shared FMD narrative running across individual stories, the telling is always situated and personally reflective. It is then told again through the actions of the researcher. This then is a co-ethnography of the UK 2001 FMD disaster.

3
Of Humans and Animals

Introduction

In May 2003, a small group of people gathered on a bleak windy former airfield site to commemorate the slaughter and burial of more than one million farm animals. Some 18,000 animals were slaughtered daily at the Great Orton disposal site. Some of these animals were infected with the virus but most were healthy animals 'taken out' in the cull of contiguous premises, or as 'dangerous contacts'. The ceremony was an attempt at regeneration and healing; Great Orton became Watchtree Nature Reserve. After the epidemic was over, landscaping work and new planting was carried out on the site and during the ceremony a huge granite boulder, a glacial erratic originally from Criffel just over the Solway, which was unearthed during civil engineering works to prepare the site for mass disposal, was dedicated as a memorial stone. Humans came along late in the life story of this boulder, but when we did, we did so in a terrible way. The plaque on the memorial says:

The Watchtree Stone

Taken from this ground
A Symbol

To the birth of
Watchtree Nature Reserve.
Dedicated on this day the 7th May 2003,
the second anniversary of the final burial

A Memorial

To 448,508 sheep, 12,085 cattle, 5,719 pigs
buried here during the Foot and Mouth outbreak
of 2001

"The tree of everlasting life takes the goodness
from the soil to sustain new beginnings"

It was an extraordinary ceremony. Here was a group of humans gathered in dedication to the memory of so many animals. Here was a strange form of grieving, but not of an anthropomorphic nature. Rather the grief being expressed had something to do with the relationship between the herds and stock buried in mass graves in containment cells and their owners, breeders, guardians. It was as if something of the people had been buried with the animals. *I couldn't go there; I could never be there, because my father's sheep are in there*, we were told at a meeting when we showed pictures of the ceremony to diarists. Alongside the need to give expression to a very particular sense of bereavement for the animals, the ceremony was also intended to be a focus for the re-dedication of the land, the place, which had been altered forever. Following a period of consultation with the local community, it was decided that the site, first an airfield, then a mass burial site, should become a nature reserve, and a competition was held for local schools to design a logo for Watchtree Nature Reserve (Figure 3.1). Fittingly, the winning entry was designed by a primary school pupil (*Emma Carter*), whose pet lambs had been taken in the cull.[37]

Farming and the ordering of nature

Contemporary views of the countryside owe much to the Romantic movement of the late 18th and 19th centuries. From the dramatic landscapes of

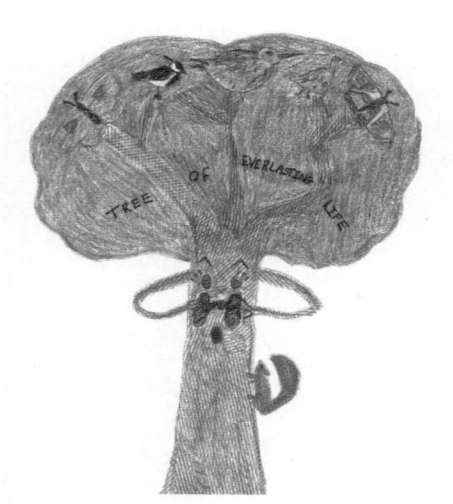

Figure 3.1 The Tree of Everlasting Life (courtesy of Watchtree Nature Reserve)

J.M.W. Turner, the romantic countryside of Samuel Palmer, the idealized pastoral idyll cherished by the wealthy clients of Thomas Gainsborough and the stylized livestock portraits of Henry Strafford, Thomas Weaver and A.M. Gauci, a notion of what is literally picturesque became inextricably linked to the places, people and livestock of farming. A visual appreciation of the English countryside is accompanied by other sensual pleasures: the smell of honeysuckle in a hedgerow, clean air to breathe, the sound of water and birdsong, the feel of moss and rock and close-cropped turf. That version of landscape, something therapeutic and pleasing to the senses, a version whose connections to the meat on the plate have become fuzzy and unspoken, was brutally destroyed by FMD. It not only turned iconic landscapes in Devon and Cumbria into taboo places of slaughter, burial and burning but revealed both a scale of farming and certain practices that people found disturbing and distasteful. It brought to light the wholesale transport of large numbers of animals by farmers and dealers. Later in the epidemic, when the disposal site at Great Orton was opened, there were television images of huge numbers of sheep being loaded, direct from slaughter, onto conveyer belts to be dumped in vast pits in an unprecedented disposal operation following killing on an industrial scale.

Alongside this animal holocaust the sensory experience of landscape itself was being transformed. The fresh air was replaced by filthy smoke from burning carcasses: witnesses told of the sky being darkened and likened the sight of flames and heavy black smoke to the burning oil wells following the Iraqi invasion of Kuwait and the first Gulf War. Local residents graphically described the smog that hung over Cumbrian villages for weeks. In the Penrith area the trees and hedges that line the network of narrow lanes were black with greasy soot. At a time of year when the countryside was usually alive with new lambs, the fields were empty, and neighbours witnessed instead the disturbing sight of heaps of dead stock waiting to be collected. The everyday noises of the farmyard, the mechanical rhythms of milking, feeding, cleaning and the social animal calls of cattle and sheep were extinguished. Those who experienced culls reported that birdsong ceased and people who lived in quiet areas away from main roads recalled being able to hear shots in the distance and the metallic scrape of loaders on concrete as they shovelled up carcasses. And when the culling was over and the stock removed, the farming families themselves were left in their silent eviscerated farmsteads with their dogs and their uncertain futures.

There are few occupations which are as complex as farming in its relationship to the wider population. We prefer in general not to think or to know too much about the processes between farm gate and plate. But livestock farming in Cumbria (and uplands in general) is not just about food production, it is integral to the character of the landscape which is in turn the basis for a large tourist economy. There are ways too in which livestock are part of farmers' lifescape in a domestic and cultural sense as well as financial and occupational. Many farmers keep photographs of their stock on the walls, particularly if it is prize-winning when there will also be trophies and rosettes on display. They may collect figurines in china or resin of the particular breed of stock they favour. In March 2002 a farmer who had kept his stock through 2001 records in his diary:

> *The photographer rang to say that she would be able to come on Saturday to take pictures of the cows. It's something that we do, take two or three pictures of the best cows just for posterity so we spent most of the day clipping cows.*

The annual agricultural shows which agricultural societies in the late 19th century developed as part of drive to improve the efficiency and productivity of agriculture, are a testament to the pride and importance farmers attach to their livestock and the interest they have in observing the achievements of their peers, sharing 'local knowledge' and discussing what innovations are available, even if they may only 'window shop' at the manufacturers' displays of new vehicles and machinery.

Farmers' lives are a very particular mix of isolation/privacy and public scrutiny. Farming is also one of the most stressful and dangerous occupations.[38] Farm families must usually cope with chronically stressful circumstances, such as economic uncertainty, low cash flow, lack of separation of family and work life, family members working in close proximity, long hours, and unpredictable weather (Hall *et al.*, 2004) A farming family many miles from the nearest centre of population may have hundreds, even thousands, of visitors a year using a right of way through their land to a local beauty spot, strangers whose names they will never know. Stock reared on the open fells of the Lake District are sold at public auction watched by a knowledgeable and critical crowd who note their condition and the price they make. A good sale before a jury of peers can be an important affirmation for someone who works alone in a remote place, as a farmer recorded in his diary going to the opening sale after FMD at his local auction: *Went to the opening sale, we did very well. Got 1st and 2nd prize in heifers, the first tickets to be given. I'm chuffed to bits, it's something I'll not forget.*

But the auctions themselves had been changed from important places of social contact and trade where farmers felt safe and relaxed into sites of potential infection and danger. The same farmer writes later in the year:

> *(My wife) asked me why I wasn't going to Carlisle fat show. I told her that I couldn't stand all day. She didn't believe that and reminded me what I used to be like. I told her I was scared to go in case I bring anything [disease] home with me. Since F&M I've been very wary, I didn't realize how much until now.*

Another farmer, after mentioning the depressing state of the weather in the summer of 2002 remarks in his diary:

> *I have only been at one auction since they opened up again, I just have no interest going when you can't wander through the pens without dressing up with oilskins on. Farming at the minute is like living in a police state. They are doing their best to put the auction marts out of business.*

A life which is a mixture of independence: no clocking on or off, no human-imposed timetable, no mission statements, is also at the mercy of weather, geography, world markets and trade agreements, food fashion and the imperatives of livestock husbandry, over none of which the farmer has any control. For example, cows have to be milked at regular times twice a day or they get mastitis, calving cows may need to be monitored in the middle of the night, and calving or lambing time requires round the clock attention. Orphan lambs need regular feeding or their mechanical feeder topping up, cows need watching to see if they are bulling. A farmer strolling through livestock is always open, sometimes unconsciously, to signals which indicate the health of his or her stock and its current position in the cycle of growth and reproduction. Put simply; livestock farming is not about just putting animals in a field to graze.

With FMD came a shift in power relations and a complete cessation of the norms of trade. When eventually they were allowed to sell stock direct to slaughter, farmers complained that without the moderation of the auction system the end-users could name their own price and took advantage to pay far less than pre-FMD, and that this price was unknown until the cheque arrived from the abattoir, a situation which again pertained in the wake of the 2007 outbreak in Surrey. While the standstill was in place older fattening stock crossed the over 30 months age barrier and became caught by the Over Thirty Months (OTMS) rule. This regulation came in the wake of fears about Bovine Spongiform Encephalopathy (BSE)[39] in cattle being transferred to humans in the form of variant Creutzfeldt-Jakob Disease (CJD).[40] Under the regulations no bovine animal over the age of 30 months was allowed to enter the human food chain. This had already had an effect on growers of breeds of cattle that were traditionally killed for meat at a later age and also on owners of dairy and suckler cows that had ceased breeding but which used formerly to be sold into the meat market. The BSE regime originally paid a set price regardless of weight or condition, this was later changed to a variable but capped rate.

However in 2001 there was a problem for farmers whose prime beef animals had been stuck on the farms at the point where they passed 30 months and they changed, transformed overnight from an important contribution to the food chain, and the farmer's income, to a waste product. All OTMS cattle were taken from the abattoirs to rendering plants. On top of this shift in business dynamics came changes in the way farmers could lead their lives. Every day there were new regulations and changes to those regulations, licences to be applied for, monitoring visits from veterinary inspectors, bio-security to be maintained. For those who had not been culled there was the ever-present worry that their area might be the next to be affected or that the infection might return, as happened for instance near Tirril, close to the site of the second case of FMD in the county in early March 2001 and where an outbreak took out a large dairy farm in late June, which came as a great shock for local people. Farmers felt invaded by the many officials who came to check paperwork and organize culls and their aftermath.

> *In November time I was about to go and get the kids from school and I thought 'nobody has rung my doorbell today'. And that was the first time since March, that there hadn't been a stranger on the farm, for some reason or another, I was sick of, sick of the invasion of privacy I think. And they would phone up and say that they were coming out on a weekend, or it would be a certain night, about something quite important like the cleaning of the milking parlour walls or whatever so you couldn't say 'oh I'm sorry that's not very convenient.*

Once the culls were over there were inspections of cleaning and disinfection procedures, and the often frustratingly slow progress back to normality. Milking equipment had to be dismantled for thorough disinfection and the buildings 'signed off', passed as safe before utilities could be re-installed and equipment rebuilt. The following diary entries from a farmer, nine months after their farm was culled, tell of these frustrations.

> January 2002: *Still waiting for DEFRA to come and check farm electrics and put (milking) parlour back together. Rang them again and again – same old answer – you are on the list.*
>
> The following week: *Another week of wondering and waiting to see if we will get the milking parlour back together. Did a lot of telephoning only to be told yet again that we are 'on the list'. Farmers are used to being their own bosses and doing what they think when they get up. Now we have to wait for DEFRA officials to come and say what can be done and when. Eventually we are told that we can put it together again ourselves, as their teams cannot cope. If we had been told this earlier then the job would have been done and finished now.*
>
> Two weeks later: *Had telephone call from where we are getting new herd from, when will we want the cows? Rang DEFRA and they say they will send the re-stocking officer (again) to check progress. Still waiting for DEFRA to send electricians. Explained this to re-stocking officer and he seemed genuine and said he would see what he could do. Switch on bulk tank seems to be faulty, sent for repair man (he) says it has been corroded* [by disinfectant used in the cleaning and disinfection period) *asked him to fit new one and check others. Re-stocking officer comes and checks paperwork. We are told we now need ear numbers of all cattle*[41] *that we are buying. Rang the owners and asked them to send them, as soon as possible. At least we have positive steps being taken. Provisionally booked lorries to transport the cattle. DEFRA re-stocking officer thinks the application could be processed in four or five days.*

The television coverage of the 2007 FMD outbreaks highlighted a tension which lies at the heart of the nexus of misconceptions between farmers and the public. A farmer broke down as he talked of the slaughter of his animals; the next pictures were of live animals being loaded into a wagon while the commentator spoke of 'good news' for farmers who were now allowed to send their stock off the farm as long as they were going 'direct to slaughter'. It is difficult to explain the difference between the wholesale culling and the slaughter for a purpose, but it lies at the heart of the concept of farming lifescape.

The sentiment outlined above from the 2007 outbreak applies equally to Cumbrian farms during 2001. Sarah Franklin (2001) speaks of 'images of farmers weeping beside pyres of their culled livestock' during 2001 and how this raised a 'number of significant moral and ethical issues regarding the relationship between farmers and livestock.' Some writers go as far as to contest that farmers who wept at the slaughter of their stock were 'simply hypocritical' (Smith, 2002). We would argue, however, that the distress displayed reflects a severe and often poorly understood disruption to a complex lifescape. Anderson (1997:119) in her critique of animal domestication, alludes to this when she speaks of the complexity of relations of domestication where 'animals can be beloved companions or eaten for a meal.'

There are, however, certainly other forms of farming where there is not the same degree of attachment or longevity. The chief difference is that outside industrial-scale farming (for example, broiler chicken units or egg-laying poultry units where large cohorts of animals may be culled at a time) it is almost unknown for all the stock on a farm to be killed at once. In a breeding flock of sheep ewes will be kept for up to six years or even longer, brought on (or bought in) in batches to give a spread of age across a flock and winnowed out as age, infertility or infirmity catches up with them. Similarly a dairy farmer will hope to keep the cattle in his/her herd for a long, economically productive life, and whilst life-expectations vary tremendously according to breed, (for instance large Holstein milk cows have a relatively short lifespan, see Figure 3.2) there will still be a spread of age across the herd. Dairy farmers will often speak of cow families (bloodline) where the young females from the best cows will come into the herd in their time and improve the physique/conformation and productivity.

In Cumbrian beef suckler herds cows typically average seven calves each and are sent for slaughter at around age ten years, a few may carry on into their teens if they are still fit and still getting calves. Just as heft (heafed) flocks on the open fells pass their flock 'map' to their progeny and appear to acquire a homing instinct, so, on a smaller scale, stock enclosed on farms will learn the geography of their surroundings: the location of gates, water, routes to buildings, shade and shelter. The livestock themselves thus form part of a farm's continuum.

The importance of continuity among livestock was perhaps not recognised until restocking began after the epidemic was declared over. Farmers told us of how newly stocked animals 'did not know their way' round the fields – it became more labour intensive to move them from field to field, there were problems acclimatising them to their new milking parlours and winter housing. We heard how they had not acquired immunity to local health challenges when they came from other areas, or how they brought new diseases with them, TB (bovine tuberculosis) became a problem on some re-stocked farms. There were problems too in patterns of reproduction where farms had not had access to a bull at the right time for a particular cycle of rearing and feeding, or where the only bull available was unsuitable and caused consequent problems with over-large calves resulting in deaths and injuries. An AI (artificial insemination) company reported that during 2001 they were unable to visit farms for fear of cross-contamination. As a result farmers were going on courses to learn how to AI cows and thus in the short term the company's income dropped catastrophically. A year later the nature of the business had radically changed as sales of semen direct to farmers took over from the practical on-farm AI service.

Humans [*and*] animals

Social scientists have long been interested in human–animal relations. Much of the early work in the 1950s and early 1960s considered livestock animals as economic units or indicators of human 'development' (Yarwood & Evans, 2000). Since the mid-1990s, social scientists (mainly geographers) have been concerned with spatial variations in human–animal relations,[42] with exploring the ways particular categories like 'livestock', 'domestic', 'nature' and indeed 'human' and 'animal' are socially constructed[43] and thus with relationships between human agency, animal agency and landscape. Philo & Wolch (1998) go so far as to suggest that this emphasis on the socio-spatial place of animals and the coexistence of and social interaction between, humans and animals, reflects a new cultural, 'animal geography'.[44]

The relationship between farmers and their livestock is neither simple nor static. Herds of sheep and cattle are not only a source of income but an ongoing part of the lives of farmers and their families. This may appear paradoxical; rearing an animal to the best of your ability, keeping it alive and thriving in order to kill and eat it – at the right *time* and in the right *place*. Farmers have different ways of coping with this and will find themselves (emotionally and attitudinally) somewhere along a continuum of care to death at different stages of an animal's life.

The emotional connection between humans and animals that are raised for food has been described as immensely respectful and devoted, even 'sacred' (Hall *et al.*, 2004). Wolch and Emel (1998) contend that culturally orientated studies of animal–human relationships highlight complex and contradictory processes of, on the one hand, consumers distancing themselves from animals as food (people eat 'meat' not 'animals' and would generally prefer not to know about the muck and blood involved in meat production) – hence an artificial split between the conceptual and the material – and the central role of animals in the structuring of society and hence to formations of human identities. Such contradictions are right there at the core of things, farming is simply about taking advantage of reproduction and growth, learning how to steer and improve those processes, whether with plants or animals, in order to produce food.

The ambivalent position of the domesticated farm animal is further emphasized by Philo (1992) and Yarwood & Evans (1998), who argue that for some, farm animals are anthropomorphic creatures constructed by the rural heritage industry (or perhaps also the romanticized view of farming portrayed by adverts for dairy products). To others, they represent an important aspect of local and rural identity, occupying a key position within the geographical imagining of the countryside (see for example, Halfacree, 1995). The 'sanitization' of livestock animals highlighted by Yarwood and Evans, where animals are clean, healthy and docile and even have pet names, may be contrasted with the violent, industrialised and anonymous death many farm animals encounter in the abattoir (Smith, 2002; Midgley, 1983).

Holloway (2001: 294) asserts that whilst animals are traditionally viewed as part of 'nature', 'the frontier between culture and nature increasingly drifts, animal bodies flank this moving line. It is upon animal bodies that the struggle for naming what is human (is) taking place'. Holloway offers a culturally informed examination of 'hobby-farming' (small-scale, part-time, food production) wherein there are emotional and ethical entanglements of human–animal relations. Animals may be viewed as friends, pets and as sources of food and as central to forming farming identities through such social practices as attending auction marts. Moreover, the animals were engaged in an ethical relation which involved regarding them as individuals while focusing on, from a human perspective, their wellbeing, happiness and 'freedom' to express 'natural' behaviour. At the same time, this relationship allowed the animals to be used at the convenience of humans for food, and to be subjected to many aspects of conventional agricultural management (Holloway, 2001:304).

However, we would argue that in order to foster a culturally sensitive rural geography, we need to move further beyond human–animal dualism to consider locally specific interdependent, fluid and shifting relations that signify how and why everything within the 'rural' is socially constructed (Murdoch & Pratt, 1993). Recent work focusing on the social construction of human–animal relations has emphasized the significance of 'setting' (Franklin, 2007), which, in a rural (English) context at least, is very much a cultural landscape, a product of human management of one form or another.[45] Thus whilst upland areas in England contain most of the remaining semi-natural habitats in the country, which often contain vegetation communities that are similar to natural communities in structure and function, their distribution is the product of thousands of years of human activity.[46] Indeed, for many people, farmland in the form of grazing land, commons, woodlands, is what 'constructs' their view of nature. As we have already illustrated in this chapter, the 2001 crisis made us aware of how the emotional geographies of livestock farming are entangled within human constructions of nature, with human and non-human identities shaped through ideas and practices played out in different contexts at different times and places. Emotional geographies of livestock-farming relations illustrate the complex socio-spatial dynamics of being someone (human) in this world. The scale of the 2001 crisis also raised awareness of how difficult it is to articulate these relations, and how these complexities, while 'known' implicitly have rarely been examined by the people for whom they are fundamental.

Death in the wrong place

As we have already discussed in Chapter 2, the scale of animal slaughter in Cumbria was unprecedented. Death became an industrialized process, with animals moving on from slaughter to containment trenches. The magnitude of animal death is reflected in the narratives of respondents.

> *Just the sheer waste, all the dead animals really, there were 2000 sheep on one of the places we were, and a lot of the ewes were pregnant and as they were being slaughtered they were having premature births and I wasn't physically sick or anything but you felt you were going to be, but you knew you had to get the job done and get on with it...it just seemed such a waste. I know stock gets slaughtered, we eat it, a lot of farm animals get slaughtered every year normally anyway – but not on that scale.*

For some, this scale of killing also impacted on their sense of identity and on their everyday living and working relations with the landscape, with livestock and with others in their community. Indeed, for many farmers the experience of farming represents a core component of their sense of identity, when livestock is removed, this core is lost (Hood & Seedsman, 2004). In Cumbria, many farmers and their families drew self-esteem and well-being from their 'day-to-day' environment. Such tacit affirmation from everyday places and sense of identity changed rapidly and dramatically during FMD. *I don't want to get big-headed or owt, but I used to feel that we were on a different planet to everyone else...[after FMD] I feel like a second-class citizen.*

As Franklin (2001:3) notes, 'the outbreak of FMD in the midst of lambing season has meant that for many farmers the anticipated period of seeing their flocks spring to life instead has seen them put to death'. There was a clear breach of normal relations – whilst lambs bred for meat are normally slaughtered, this is not when they are newborns, and so the rhythm and cycle of livestock–farming relations was out of synchronization. The epidemic created fissures in taken-for granted lifescapes which transcended the loss of the material (i.e. livestock) and become also the loss of the self, calling into question perceptions of identity and meaning. Death was not only in the wrong *place* (the farm rather than the abattoir), but it was also at the wrong *time* (in relation to the farm calendar) and on the wrong *scale* (such large scale slaughter seldom occurs in one place at one time). As a slaughterman pointedly noted *farms are made for farming...and slaughterhouses are made for slaughtering cows and farms are made for farming cattle.*

> *The worst thing was when they started bringing trailer loads of newborn lambs in. That was terrible, we had to go in and unload them but it had to be done. We unloaded the trailer and drove them into a pen and got out of there as quick as possible because you can hear them, the animals bleating as they were being herded up to be put down. Absolutely innocent young lives.*
>
> *We are outsiders to Cumbria. If anybody had told me I could feel this way about such a thing, I would have found it hard to comprehend. The emotions generated were so strong even amongst those of us who were not farmers. I could not tell of the images in my mind of the dead cows immediately over my garden wall. The sight of them lying there, the smell, the vision of them being lifted up by tractors and piled into lorries. I remember thinking 'how many more lorries? How much more cleaning? How can these poor people cope with this?' We are all going through bereavement and shared the feelings of grief.*

> Like you go to a slaughterhouse everything's set up....You can't make it on a farm eh, not when you're expected to go two minutes, set up, ready, you just can't do it eh... I dunno. It just sort of got to me like. You used to go to farms and grown men used to come and cry like.

The culls brought for some a sense of failure, a loss of professionalism and a sense of not having 'done the job right'. Holloway's (2001) emphasis on good 'stockmanship' (sic.) is borne out by many amongst the study farming community. Stockmen speak of pride and satisfaction in relation to the process of rearing healthy stock. There are clearly economic benefits associated with this, stock that are 'kenned'[47] are more likely to thrive and thus command a higher price at market. This pride can continue once the animal has changed from being on the hoof to being a carcass, from being an individual animal to a commodity. For example, some farmers might telephone an abattoir to find out 'how a beast has killed'[48] at a small abattoir they might even go to view the carcass[49]. Here a farmer discusses his interest in how one of his bulls 'has killed': *We had some bulls to go to the abattoir they killed out very well, we had a good trade.* Another farmer discussed contacting the abattoir after the slaughter of his livestock: *Phoned (abattoir) when I got home to find out how our bulls had killed.* Again, Holloway (2001) notes the ambiguity of this relationship, and quotes one of his study respondents, who when taking an animal to the abattoir asked the slaughterman to 'look after her babies'.

Drawing on the work of Wolch & Emel (1998), whose review of culturally orientated studies of animal–human relationships highlights complex and contradictory processes of animals simultaneously being viewed as friends and as sources of food, and Anderson (1997), whose critique of domestication suggests that taming and regulation of animals may help humans to construct a sense of superiority, we suggest that when it comes to farming there is a deep and complex bond which stems from a much more practical and pragmatic working relationship between livestock and their handlers. This relationship stems from the lifescape, the mutually constitutive interrelationship between people, place and production system (Ingold, 1992; Howorth, 1999; Convery, 2004), which we explore further in Chapter 7. In Cumbria, this relationship is epitomised by the *heft*.

Animal ways of knowing: hefting

Hefting refers to the ability of certain breeds of sheep (in particular Scottish Blackface, Kendal Rough, Herdwick and Swaledale) to maintain a home range group structure. This is essential in areas of common grazing where there are frequently no physical boundaries between the flocks on one farm and another. Research into hefting instincts in sheep has demonstrated that hill sheep follow particular movement patterns which are probably an elaboration of a movement pattern adopted from her mother in the first summer of life and further, that once hefted properly, the sheep will stay within a common area and tend not to mix with sheep from adjacent hefts. This system – which is essential for the traditional management of upland grazing areas, and which has in turn shaped the characteristic flora and appearance of the uplands – is dependent on the knowledge of range and the 'virtual' boundaries with other groups of hefted sheep being passed from one generation of sheep to the next. In essentially 'teaching' each other the location of the heft and largely managing themselves in terms of their geographical distribution, hefted sheep are an enormous economic asset to upland farmers and important for the maintaining of traditional farming (Burton et al., 2005:33). They are an iconic and important facet of the Cumbrian agricultural landscape[50]. However there was disruption to the system when parts of commons were cleared when their flocks were culled in 2001 and stocking density and coverage was drastically reduced. The surviving sheep, without the benign pressure of neighbouring flocks to keep them 'in their place', began to drift over larger areas and into new territory causing shepherding problems. In some areas disputes arose as to whether fencing should be erected across commons to keep sheep to their traditional hefts, and to encourage the new restocked sheep to learn where they should be.

Moreover, permanent herds and flocks are in effect hefted within the farm itself, that is they possess unique knowledge about the geography and routines of the farm (Briggs & Briggs, 1980). It is rare for a farm to replace all its livestock at once, thus the mass culls of 2001 signified also a loss of knowledge of complete herds and flocks of livestock. It is customary for livestock to be replaced on a rolling programme, thus there is a continuity of knowledge passed on by older members of the flock/herd who know the geography of the farm (fields and buildings) and will know where to drink, where to eat, where the shelters are and which gate a dog will shepherd them to when it sets off round the field. The loss of the hefted flocks, however, represented much more than the loss of livestock 'knowledge'. As Franklin (2007) notes, sheep genealogies mix some of the oldest traditions of belonging to blood, soil and country. The heft is bound up in the genealogy of the farm and the land; it is part of the farming lifescape. The farmer and the sheep both articulate ways of being and doing developed through lived practice (concepts developed further in Chapter 7).

After re-stocking, farmers spoke of flocks meandering about when they do not know the ground and sheep dogs and humans having to work much harder to control their movements:

> *It* [lambing] *could be a harder time this year as none of the sheep have lambed before. It is easier when there are older ones for the first-timers to follow their example. Even feeding them at a trough takes time because they haven't done it before and there is no older ones to teach them.*

So farmers take pride in 'healthy', 'well-bred' stock that are 'kenned' on the hoof but that will nevertheless bring economic profit as a carcass. At the same time and as we illustrate below, whilst the slaughtering of livestock is part of livestock–farming lifescapes, few farmers and livestock handlers would volunteer to be at the interface between live animal and food. The abattoir would normally provide spatial distancing and some emotional detachment. The scale of killing during the 2001 crisis transgressed this emotional geography of farm as the appropriate place of livestock management and the abattoir as the appropriate place of livestock death. The culling regimes covered key spaces on the farm and parts of the farming landscape with death and dying:

> *And he'd had them*[sheep] *out on his fields and he'd gone with tarpaulins and bales of straw and making them little places to shelter so if the weather was cold or wet they could take their lambs inside, because I mean the sheep do have a bit of sense and they'll get inside and get them, the lambs, keep the lambs dry and warm. And he'd done all this and he'd really worked and worked and worked and I went past one day at the weekend I was on my way down to my daughter's and there was two heaps and they were just lying by the side of the road waiting to be collected one day, and I cried then (upset here) and I see the farmer's wife sitting crying and the farmer sitting crying and we were just wondering what on earth's going to go on?*
>
> *Like I said for most farmers it was a traumatic experience seeing their stock slaughtered. A lot of them couldn't face seeing it I know some camped out for the night but most farmers on the farms I was on were actually involved in helping to slaughter the stock. They wanted the team to work with them. They felt the need to be there they wanted to make sure things were done correctly.*

Emotional geographies of changed lifescapes

According to Humphrey (1995: 478), humans 'can hold multiple, seemingly contradictory attitudes to the same animal: enslave, worship, consume, abuse, befriend, hunt, play games with, grieve for'. Figure 3.2 presents diary and interview accounts relating to Holstein–Friesian dairy cattle, together with other documentary sources, to demonstrate how perspective and discourse portray the same animal as a 'machine', a 'friend', a 'representation of a life's work', an ephemeral presence, or a 'bovine replicant' with limited lifespan (after Dick, 1968).

As we mentioned above, the emotional geographies of livestock-farming relations are embedded in the production of particular places where such relations are played out. The intricacy of this relationship meant that for some, the cull of dairy cattle signified the total severing of a whole array of networks of meaning, practice and identity:

> They all had names here, they weren't just numbers, they all had names and known individually. They are all characters and individuals in a way when you farm the way we do.

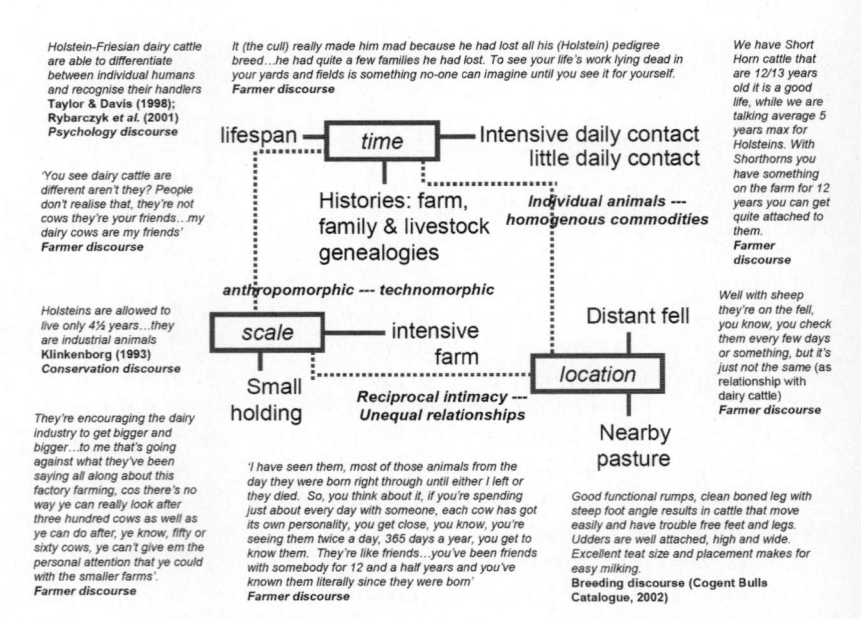

Figure 3.2 Complexities of livestock–farmer relationships: the case of Holstein–Friesian dairy cattle

> They walked them through the milking parlour and out, through the milking parlours like, as if they were going to get milked. After they come out of the parlour we'd get about ten through, close the door off and they would just stand there and be shot. They weren't a bit bothered. Dragged out and put on the silage pit and another twelve through, and that was just like a conveyor belt really. It was just like milking them or something. There was about 30 cows due to calve in the next month, and they were all in two calving sheds, calving bays, and they were just done where they stood, and dragged out. And there was a calf, and the slaughtermen had had to draw straws on who shot the calf and they were absolutely devastated.

A further level of meaning and identity can be linked to the cull process itself. The culling regime imposed by DEFRA during the crisis undoubtedly meant that a significant number of healthy animals were killed. As the National Trust (2002) indicates, animals of particular significance, including pedigree bloodlines, sheep managed by hefting and livestock associated with distinct localities were lost at this time.

> But I mean I just keep thinking of the farmers round the corner that had a pedigree dairy which were distinctive even to me, they were brown and white cattle, very beautiful cattle. And they, I think they went back 160 years the pedigree. But they were all gone.

> We were lucky that we had 40 cattle up at another farm, they still had the bloodlines of course when they came back, they were very poor and lean but we still have the bloodline.

> To see your life's work lying dead in your yards and fields is something no-one can imagine until you see it for yourself.

'Life's work' is a point often missed in the official inquiries and in how the crisis was understood by the urban and political classes at the time. It refers to the intimate knowledge built up over successive generations of breeding stock, a closeness with the livestock in which animal human lives are lived in a mutual space and often stock lines relate to family members:

> H: But we both had, cow families from when we...
> M: Before we were married, like really eh, from both our parents farms, you see when we started we sort of got something with each parent

> so unfortunately those are the bloodlines you cannot replace at all you see.
>
> H: I was given one when I was seven...

'Life's work' lies dead in 'your yards and fields': here a sense of place is mutually shaped between the livestock, the herd which belongs in that place (yet has been slaughtered), and the physical spaces on the farm, rendering that place a site of bereavement (Figure 3.3). This farmer shows that the nature of the livestock/human/place relationship is also an emotional one, and its fracturing is profoundly shocking and cannot be imagined. The blow of seeing this graphic destruction of 'life's work' creates a division between those who have been there, and those who have not.[51]

As discussed in Chapter 1, this sense of a chasm opening up between the insider and the outsider is well known in disaster studies (Erikson 2003). The diaries and interviews in the Lancaster study, and later the archiving project which followed, came to serve as a conduit between different ways of understanding the impact of the disaster. Having the opportunity to tell their story and have their experience recognised,

Figure 3.3 A life's work

served in some way to reduce the gulf which had opened up between those that suffered some of the worst horrors of 2001 and those who had not.

FMD also served to reinforce rural identity, particularly in the context of a perceived urban and remote central government. Bickerstaff *et al.* (2006:855) note a shared sense of place that 'reinforces within local culture the unequal political–economic relationships of centre and periphery...a common structure of feeling is reaffirmed and sharpened under conditions of domination, such as those associated with the government handling of the FMD crisis.' This feeds into a general loss of trust in governance and policy, a feeling summed up by this comment from a farmer post-FMD:

> *They have no credibility at all in my eyes now, because I know what they used to say to the press in London and what was actually going on in reality up here, and you know it's very sad that you've lost* [trust] *of* [government] *always being there in a crisis once it let you down badly.*

Here another farmer questions public goods provision, for some time an important plank of British agricultural policy.[52]

> *Country people have come to the conclusion that they have to keep the countryside spick and span for the townspeople to enjoy on a Sunday. The government don't seem to realise what kind of effect they are having on country life. I think farmers are going to have to struggle on as we have for the last few years. It makes you wonder if it is all worth it.*

Summary

The 2001 FMD crisis severely disrupted the tangible, material and tactile relationship with known livestock. As Humphrey (1995:478) puts it, 'we can hold multiple, even seemingly contradictory attitudes to the very same animal '. We argue, however, that the complexities of this relationship have not been adequately reflected in either the public or academic debates about livestock– farmer relationships during the FMD crisis.

The events of 2001 transcended the loss of the material (traumatic though this undoubtedly was) and became also the loss of the conceptual (the loss of the meanings associated with this way of life). The scale of killing during the 2001 FMD epidemic did transgress the emotional geographies of the farm as the place of livestock management and the abattoir as the place of livestock death, because death was in the wrong place, at the wrong time and on the wrong scale. Like Hall *et al.* (2004), we consider that a deeper understanding of the nature of humans' attachment to their animals and the meaning of this relationship in different sociocultural and occupational groups has enormous practical implications for (animal) disaster management. As one farmer poignantly notes: *There's not a day goes by as I don't think of it ... Big thing like that I suppose it don't just stop, does it?*

4
Things/Materials

Introduction

In accounts of disasters it is often the case that everyday objects, materials, tools and spaces take on heightened significance, become transformed or enlarged into resonant materials which carry traumatic associations, even agency. This kind of transformation can be traced in so many accounts from the 2001 FMD crisis but also in very different disasters, for example the great Japanese earthquake of 1995. In his ethnographic study Takashi Harada (2000) recalls how he worked as a volunteer in a relief centre and recounts that upon seeing an elderly woman struggling with heavy bags near the railway station he offers to help her carry them. The elderly woman was on one of many exhausting trips to carry dishes which she had gone back to rescue from her earthquake damaged house. Later Harada ponders about what made her make so many hard journeys bearing heavy loads on the congested trains and buses. He suggests that for her these represented a strong desire for continuity in her daily life. The old woman had herself survived a disaster, but at the same time had had to abandon the place where she had led her life. She needed some things which had belonged to that life before the earthquake; the will and the act of bringing back belongings demonstrates this desire for a little continuity.

Harada's discussion is influenced by the actor-network approach developed by sociologists, notably Bruno Latour, Michel Callon and John Law. In this approach there is little if any distinction drawn between the 'social' and the 'material', and, as a theorization of power, the actor network approach sees action as an effect of the mutual enrolment of the social and the technical. Indeed the approach is characterized by ideas about the delegation of social relations to materials and spaces. Power, as articulated by Callon (1986) and Latour (1987) and drawing on Foucault, is not a possession but an arrangement of assent, carefully ordered within an actor-network. Things and materials can embody some intimate connections and associations within the network. The network can be strong or it can also be fragile, because it takes a lot of work to keep it hanging together.[53] The network also involves a wide range of interconnections; a world in which the 'and' of the connection between actors defines the *time* and *space* of the relationship. As Hetherington & Munro (1997:197) note, places therefore become the effect of 'folding of spaces, times and materials together into complex topographical arrangements that perform a multitude of differences.' The location and role of materials within a particular social network may thus change in many different and unpredicted ways. The best way to understand this is through examples, or case studies, the favoured approach of both actor network theory (ANT) and its interdisciplinary relation, Science and Technology Studies (STS).

The development of 'non-representational theory' by geographer Nigel Thrift (1996) also provides a useful perspective on the material dimensions of human life as it seeks to integrate non-human objects of all kinds and the human world. Thrift's ideas can also be traced back to the writings of Heidegger (1971) and the notion of dwelling,[54] which Buttimer (1976:277) states implies more than to 'inhabit, to cultivate or to organise space. It means to live in a manner which is attuned to the rhythm of nature...anchored in human history and directed towards a future'.

For the Ojibwa of Northwest Ontario, one of the kindest and most reliable elements in their 'dwelling'- the living waters – had slowly become contaminated, deadly, due to the 20,000 pounds of highly toxic mercury which, it was discovered in 1970, had been working its way down the Waibigoon River from a paper and pulp industrial site some 80 miles upstream. The presence of methyl mercury, bio-accumulated in fish and other aquatic organisms upon which the Ojibwa depended on for their livelihood, had been slowly poisoning them. The subsequent closing of the rivers and the declaration of the fish as unfit for human consumption, destroyed not only the Ojibwa's income but a way of life characterized by very particular skills (Erikson, 1994).

Although generalizations are problematic, in many indigenous communities such as the Ojibwa, knowledge about the 'environment' represents the principle of sociality, as social and cultural life is reflected in their surroundings, in this case the waters within their territory (Seeland, 1997). In such communities, the waters are not an environment in a technical sense and should not be considered in terms of natural sciences, they represent a local world, synergistic *places*, where culture and society, needs and resources, the past, present and the future coincide (Convery, 2006). The waters *are* the people.

Within environmental sociology, Milbourne (2003:159) states that the work of Latour (1992) has introduced notions of hybridity to describe the ways in which the natural and the social connect within complex networks and Irwin (2001) has discussed the idea of a 'co-construction' of society and nature in order to capture 'the dual process of the social and the natural being variously constructed within environmentally related practices and particular contexts.' Irwin thus argues for the need to unravel the complex connections between nature and social and cultural life, viewing environmental constructions as projections of wider socially generated concerns and problems. Environmental problems do not sit apart from everyday life but instead are accommodated within (and help shape) the social construction of local reality. Macnaghten & Urry (1998) discuss the importance of 'embedded social practices' in understandings of nature, and the ways in which nature is constituted through a variety of socio-cultural processes. For example, Robin (2002) writing about everyday spaces in people's lives, discusses the difference between the meaning of the English term *house* and the Maya term *nah*. The term *nah* was used in the Classic Maya period (AD 250–900) and is used in a number of contemporary Maya languages. In contemporary Yucatec Maya, *nah* refers to a complex of core domestic and work spaces that are sources of knowledge about the world, including houses, yards and fields, or, in other words, sleeping spaces and habitual workspaces. It is not the structure we would refer to as a house in the West.

Wilson (2003:7) states that for Native American cultures, one of the most visual cultural symbols found on the land is the sweat lodge. In a sweat lodge, saplings are tied together to construct a dome-like structure. In the middle of the sweat lodge there is a pit in which the grandfathers (rocks) are placed. During a sweat lodge ceremony, the inside is made completely dark and water is sprinkled on the grandfathers, which creates heat. Within the sweat lodge individuals come together for praying, drumming and singing.

During the FMD disaster, familiar objects such as straw, disinfectant, tractors and low loaders, which have their place in daily work become *re-placed* as objects of biosecurity or mass slaughter. This shifting between the ordinary and the extra-ordinary, even bizarre, can later return to haunt people, since certain objects found in everyday settings have become totemic and, because they are close at hand, can become instantly evocative even if they occur in different settings. By totemic we refer to what Erikson (1995) describes as a tangible identity marker for a group. In the same way that Durkheim mentions flags as a kind of totem, things and materials associated with FMD ('no entry signs', the presence of a 'Snowie wagon' for animal carcass transportation) became totems of infection, the sense of 'being dirty', and ultimately of loss.[55]

A recent review of the 'lasting impacts' associated with floods in the UK showed that irreplaceable items and lost or damaged memorabilia were ranked as a 'major impact' (Risk and Policy Analysts, 2005). Following the flooding of the Cumbrian city of Carlisle in 2005, residents became acutely aware of lost material items, and victims spoke of the need to *'reduce the clutter'* as a form of protection against the material loss and trauma of 'future' flood events (Bailey & Convery, 2006).

More recently, research by Lucy Payne in South Yorkshire following the UK 2007 summer floods has focused on the use of personal materials and space post-disaster. She notes how some families displaced by the floods made big efforts to decorate their temporary homes for Christmas, trying to recreate the social in the smallest of places, and against the 'benign institutionalization' of their living space. Her close observation of temporary space makes visible both the incongruity of materials and their critical role in domestic space:

> The caravans appear small although they have been described to me as 'state of the art'. They are surrounded by metal fences and at the back big mud piles. Big red gas cylinders add to the sense of 'utility' and 'hazard'...The caravans are close to each other ... the way they have been positioned means that the windows do not look directly into each other.
>
> Lucy Payne, personal communication (14-01-08)

Payne's research echoes the work of Harada (2000: 208–211), who as we discussed earlier, writes about the transformation of space (public to private) following the Kobe earthquake in Japan, January 1995.

> We started to talk about the spaces as particular people's 'houses'. These small areas represented the construction of 'private space' in the 'public shelter'. So the 'public' shelter changed to a collection of private spaces: you could not freely enter a square surrounded by four boards because it belonged to a 'specific' family or person. Their belongings lay on the floor covered by a board, and at night they lay down and slept there...Gradually the shelter became a more and

more closed space...through the process of making private spaces for themselves in the shelter, the refugees and voluntary staff were able to reconstruct their individuality and 'social' context: the refugees lived and spent their time there. This suggests that 'individuality' makes sense only in a 'social' context, and the ordeal of seeking continuity in our daily life is also part of the process of individualization.

Materials then can become metaphor. Indeed, as the anthropologist Christopher Tilley (1999:271) argues, material metaphors do not just

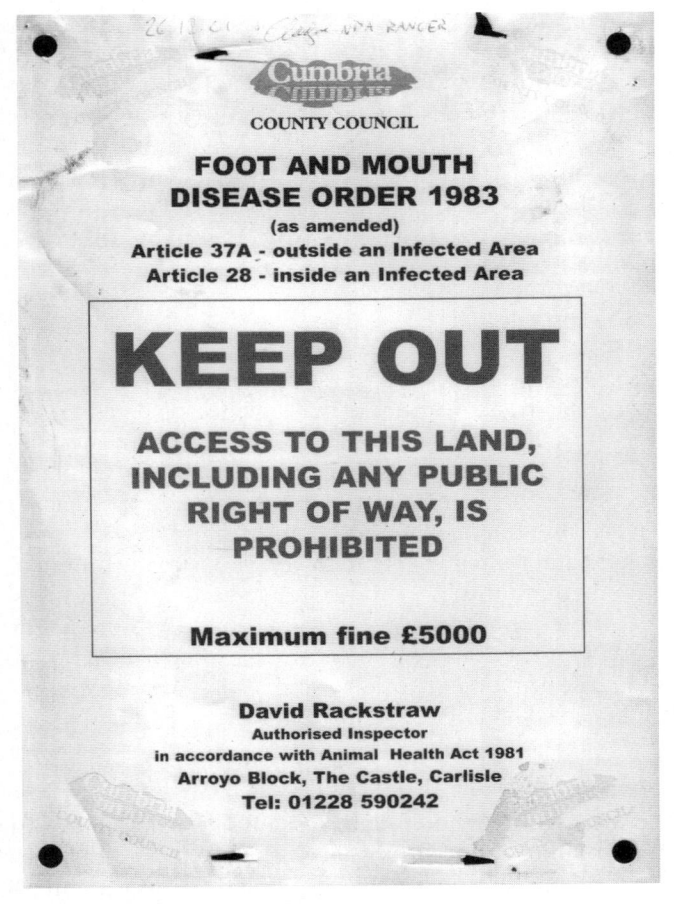

Figure 4.1 'Keep Out', Cumbria 2001

Figure 4.2 No entry signs on footpath gate, Surrey 2007

reflect social realities, they play an essential role in the description and (re)definition of those realities. Thus, images repeatedly shown on newsreels and newspaper front page splashes during the 2007 FMD outbreak in Surrey, for example depicting straw placed across the farm entrances; scenes of disinfectant being sprayed across roads around farms; loaders and tractors moving about the farmyard;'no entry' signs on farm gates and footpaths (Figures 4.1 & 4.2) shot a chilling message around rural communities in the UK. The configuration of ordinary things in an extraordinary way, remained highly resonant for those who experienced the 2001 FMD disaster, and signalled that the UK was suddenly in the grip of a major new scare. These headlines and stories are typical of the days after the 2007 outbreak:

Show goes on despite animal standstill

AS the spectre of foot and mouth cast a dark shadow across the agricultural community, country shows and livestock sales have had to make emergency arrangements following the Government's ban on the movement of cattle, sheep and pigs. Now they can only wait and hope Ministers have learned the lessons from the 2001 outbreak that will help protect their livelihoods.

(Yorkshire Post, August 6th, 2007)

Foot and mouth again

THE spectre of the foot and mouth epidemic that brought chaos to Britain six years ago returned last night when a fresh outbreak was confirmed. Prime Minister Gordon Brown cancelled his holiday and joined an emergency Government meeting after being told the disease had been found on a farm near Guildford, Surrey. Meanwhile a nationwide ban on the movement of all sheep, cattle and pigs was put in place.

(Western Mail Aug 4th 2007)

The photograph in Figure 4.3 (from the 2002 Westmorland County Agricultural Show) indicates the 'emergency arrangements' alluded to in the Yorkshire Post story above. This often meant that shows went ahead either without animals or with strict biosecurity measures.

Figure 4.3 Westmorland Show 2002

Material cultures of disasters

So here we are talking about the significance and power of 'things': materials, artefacts, events and the relationship between material order and social meaning in times of crisis. Harada (2000) argues that sociality is made up out of the material objects which people use and which surround them in daily life. Disasters can re-order meaning and 'things' can either become out of *place*, or out of *proportion* during disasters. So in this way, the continued resonance of certain objects marks out those who experienced FMD from those who didn't. For example the lines of newer tarmac, roughly 12 feet apart, which now cross many small roads in Cumbria (where matting was laid and regularly soaked with disinfectant) are probably not noticed by many drivers, but they remain a daily reminder of 2001 for local residents.

Again, remnant pieces of carpet which, when found in the house or stored in the attic are just benign or inert materials, during the 2001 epidemic became objects of biosecurity when placed across the farm gates and soaked in disinfectant. The 'Keep Out' notices and plastic tape at the entrances to infected farms were a public warning of pestilence within. Vehicles took on special significances: high sided leak-proof wagons, load-alls, elevators, low-loaders all became agents of slaughter in 2001, especially when combined to form a convoy – mass slaughter depended on particular combinations and numbers of vehicles. Snowie's was the contractor which transported more than a million culled animals to the disposal sites in Cumbria in 2001.

> One of the first times I went out [after their cull] *we decided to stop at a garage for a paper and who did we bloody bump into? It was a flipping slaughter team getting their sandwiches for the day. We were sitting in traffic lights and we were behind a Snowie wagon, and we went on the other side, there was a wagon load of pens going the other way, low loader full of loading shovels and you knew were they were all going. And we thought blooming heck you couldn't get away from it. But as summer went on you saw these Snowie wagons and these things and you knew what they were taking, were on with, who they were taking out, it but you got more used to it. The first few times you saw them the hairs on your neck used to stand on end, and you knew what they had on board, what they were doing or they were on their way back for another load.*

After the crisis died down, the sight of a Snowie wagon anywhere continued to evoke horror as four respondents wrote:

> We blood tested our sentinels on Thursday and the New Zealand lass, that is the restocking vet, turned up, and she is very pleasant lass. Then Snowie's wagon came in the yard [I said] 'what the hell is that coming for?' Oh she said, I asked them for 3 handlers to catch the sheep, they must have just come in that as their vehicle. I said 'that is the thing that carries the foot and mouth in the back I don't want that here.'
>
> We were affected with animals being taken away and I can remember one particular morning seeing you know, when I was going on my usual walk about half seven eight o'clock, having these Snowie lorries rumbling by and it just sent a shiver down the spine you know so I think that's the way it affected us.
>
> Big sale advertised in farmers weekly. Snowie's surplus equipment. Snowie's name was at the forefront of the F & M battle. His wagons carted all the dead cows and sheep away. Looking at all his equipment made me step back and think a bit. Sort of hoped that it would be the end of another chapter, however some of the wagons and loaders had been withdrawn due to the suspect case in North Yorks the other week. So it might not be over yet. Friday we went to view all the machinery that was for sale by the Snowie group tomorrow. Mainly lorries and load-alls and pressure washers. Lots of farmers and contractors talking about the prices and how much money Snowie will have made...It was quite eerie and upsetting looking around the trucks and the load-alls. And thinking about the jobs these machines had been doing this time last year.
>
> Out of curiosity Jackie went to Snowie's sale of equipment, even though the lorries, trucks etc were thoroughly sterilised they still stank of dead animals.

Spaces can become more private *and* more public at the same time. Farms for instance are at once domestic and occupational spaces and both were opened up to public scrutiny (and sometimes vilification) during 2001, so that the private became public. Biosecurity measures meant increased isolation, few or no social visitors at all, or movement on or off the farm without an appropriate licence (Figure 4.4). However at the same time, farms were cast as objects of surveillance and risk (Figures 4.5 & 4.6).

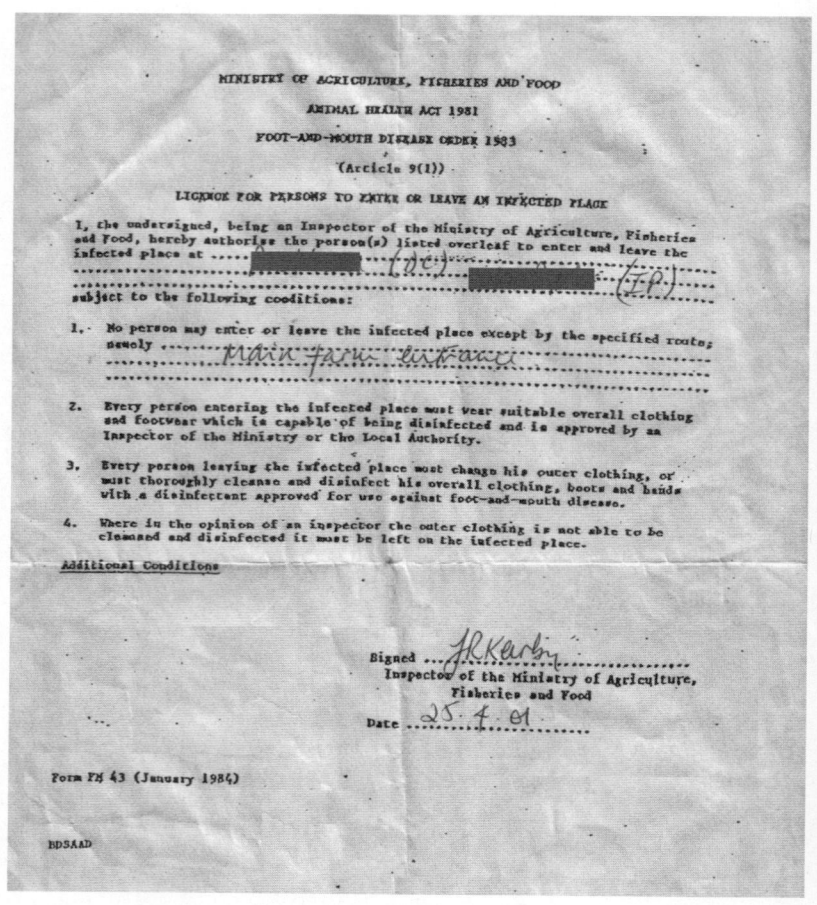

Figure 4.4 Licence to enter or leave infected premises

Cleaning & disinfection

Those farms where infection was confirmed or suspected, were classed as 'dirty' and following the cull of livestock, were legally bound to be cleaned. The complexities and costs of the cleaning and disinfection required after slaughter on 'Infected and Dangerous Contact Premises' are well documented in the National Audit Office Report (NAO, 2002). Following preliminary cleaning or 'damping down', the washing away of blood and faeces from slaughter sites, the secondary cleaning

Figure 4.5 This footpath is closed, Cumbria 2001

THIS FARM IS STILL UNDER 'A' NOTICE RESTRICTIONS

The right of way through the farmyard is still closed.

Thank you for your continued co-operation

Enquiries to DEFRA 01228 591900

Or Cumbria County Council 01228 590242

Figure 4.6 'A Notice' access restrictions

process had to be completed before re-stocking could begin. Wooden fittings from older buildings which were deemed un-cleanable were ripped out and burned, hay and straw which might have been exposed to infection had to be destroyed. There were occasional buildings, judged too fragile to withstand pressure washing, which were demolished altogether. In some historic barns and sheds the limewash render on the walls had to be picked off and the walls re-rendered once they had been cleaned. 'Cherry-picker' hoist equipment had to be hired in to enable the thorough cleaning of beams and roofs of high modern cattle sheds.

Buildings were cleaned with high-pressure water and disinfectant and this led to consequent disposal problems with many farms having to install special 'lagoons' in which the waste water was stored until it could be safely disposed of. A water company in the North East complained that it had not been consulted about the immense quantities of water needed for cleansing and disinfection and had narrowly averted a local supply 'incident' when huge volumes of water were drawn from the network. Some farmers elected to do their own cleaning and disinfection and were paid an hourly rate by DEFRA for the work. Some agricultural contractors who had lost their usual work set up cleaning teams, but the Anderson Inquiry was told that the work was sometimes undertaken by 'contractors who had little or no knowledge of farms or farming' and that 'there was no standard approach to follow because the structure of farms varies widely.'

Buildings, yards and silage pits had to be cleaned to an 'almost clinical' level of spotlessness and inspections were rigorous and unforgiving, failure to pass meant that the job must be done again. Unsurprisingly, biosecurity regimes were seen as punitive measures imposed by officials from an alien culture:

> We had to have a field officer[56] watching us cleanse and disinfect so every time from here to there and it took an hour a round journey to bring either two or three heifers, depending how many could get in the trailer. We had two bio security police sat at the top of the road here. So I think there was actually five ministry officials involved that day and us. Us and our children running around this field, like headless chickens, trying to round these heifers up. None of them, we didn't want them in our field to help, but they all just sat there watching, not a care in the world.

The costs of cleaning were also enormous, and on the 23rd July the Prime Minister ordered a freeze on all secondary cleansing and disinfection while DEFRA attempted to obtain estimates of costs for all the work in progress at that time. This caused problems for 'outside' contractors and for the farmers using them who faced more delays to plans for re-stocking and a return to normality. It also had knock-on effects on the Government's wish to re-open rights of way across farms since even where local authorities had declared paths open, the restrictions imposed on farms with 'A' notice status meant that any paths over them had to remain closed until they were finally 'signed off' as clean.[57] This confusion over the status of farms and paths led in some cases to conflict (see quotes from respondent) and in some cases farmers having to put up additional notices explaining why paths were still closed (Figure 4.6). As the NFU said in an e-mail bulletin in mid-July: 'the announcement comes at a time when many farmers are being asked to undertake stricter bio-security arrangements. Despite DEFRA advice that the risks are low from walkers, farmers perceive that their biosecurity efforts could be jeopardised by walkers moving from farm to farm'. Farmers failed to see how walkers were not a risk when they had received official advice from DEFRA one month earlier asking them to 'keep up their guard against foot and mouth' and to be 'vigilant at all times'.[58]

> A tourist information centre worker remembers that: *I had to ring DEFRA for some definitive advice on paths which didn't seem to be obtainable anywhere. DEFRA were unable to help really. They did get back to me in the end but I thought they were pretty slow at it. They must have known what the questions were and known what the answers were, if they didn't they bloody well should have.* He adds that *at one stage I wanted to put a disinfectant mat outside the Tourist Information Centre for three reasons really; first, I didn't want to get my ear bent by farmers. I'm dubious about whether it works or not but anyway, I wanted it out there to show the farmers we were on their side, second, it may have worked so lets have it. Third, I wanted to make sure that any tourists who did come realised that there was something amiss and that you know they should abide by things you know, because they come from in town and see a disinfectant mat. But the council told us we couldn't have one it was against their policy. I mean I find that unbelievable.*

Dirty farms and dirty farmers (any other workers who attended for culling were also classed as 'dirty') were not supposed to mix with 'clean' spaces or people. These categorizations were public ones, given meaning by approved regulation and attention from outside the farm, yet they affected and altered the nature of the private space.

This echoes the bureaucratic role in re-traumatization related by Erikson who writes of how the Canadian Department of Indian Affairs gradually moved the Ojibwa from their scattered river holdings to new reserves, where they would have access to modern services. The short distance move was Erikson argues, an attempt to create a break with the 'timeless world of a hunting and gathering people' and a link with modern industrial society. The Ojibwa were rehoused:

> Houses in the new reserve were rough and spare. Only a few had the promised piped-in water; fewer yet had running toilets. They looked as if they had come off an assembly line – neatly the same, neatly spaced, arrayed in symmetries like a military post. This may be an appropriate geometry for places where the land is flat, the roads straight and a kind of quadrate orderliness is admired, but it makes very little sense in this wilderness clearing where nature is a thing of rises and curves. And it is wholly alien to the native sense of patterning.
>
> (Erikson, 1994: 47–48)

How public protest stopped the pyres

Undoubtedly the building and burning of pyres during the early stages of the FMD disaster was an example of how policy itself could inflict extreme distress.

> *I suppose the first impression was a big pyre at Sockbridge* [near Penrith] *and that was a cull of 9,000 sheep and it burnt for three weeks. And the thing that struck me about the whole thing was a disease from I don't know, historical, its been around for ages, which was being dealt with technology from the Middle Ages.*

As a young farmer wrote in her diary:

> *First the slaughter then the burning, then the clean up. It was just days of culls after that. Saturday, Sunday, Monday and then the pyre was going on, it was just, I can't even say what it was like now. It was so cold as well, oh I feel really cold even talking about it and then we'd stay up like until three o'clock in the morning drinking but you never got drunk it was the weirdest thing.*

At the beginning of the outbreak, when burning was still the preferred method of carcass disposal, the 'days of culls' were followed by the building of the huge pyres, constructed using old track rails, railway sleepers and fired by vast quantities of straw, wood and coal – everyday materials, familiar on most farms, but which became symbolic of disease, destruction, horror.

But these pyres also became symbolic of policy blindness and mismanagement due to lack of relevant local knowledge. The grade quality of the coal used for the pyres became an immediate and practical issue, symptomatic for many of the handling of the disaster as a whole. As one farmer pointedly noted: *They were using crap coal in some of those (pyres) so they weren't burning properly*. A resident of Longtown told us:

> *The first pyre they built, they sent wrong coal it wouldn't burn, they sent wrong coal. It was furnace coal not stuff for open air. They were having to go and heat it and create drafts and all sorts but it just wouldn't burn. There was one farmer, he went down with foot and mouth and they spent all night putting coal in his coal shed loading it with wheel barrow and then next day as soon as they put it on pyre, it went out. An then they spent the next day, they spent all day shovelling it back out again. And it was the lads* at [company] *and they was saying this is stuff that they burn at Sellafield but DEFRA weren't listening and the mess that some of these lads were making on farms, trying to build pyres, fetching coal and then fetching it back again, you've no idea. So disorganised, so disorganised* (shakes his head and reaches for his cigarettes).

Here another resident of Longtown discusses what it was like to live so close to the pyres and how 'they' were using the wrong materials to build the pyres.

> *Well, there was five pyres all burning at the same [time] within a quarter mile radius of Longtown…I got up one morning and the smell in the house was terrible, and I thought it was foggy outside. I opened the back door thinking to let the smell out and it was worse. It was smoke swirling around the houses. Because it was a calm morning, you would think it was thick dense fog. And it was stinking, it was horrible and we had that for three or four weeks, because they used the wrong coal on the pyres. They were building the pyres wrong. They weren't burning properly and some of them pyres burned for six and seven weeks.*

A DEFRA field officer recalls having to clear up after a poorly constructed pyre: *During my last two weeks I had supervised pyre ash removal. Some of the pyres had been huge and one still had hot ash. On some pyres it had taken three days to remove all the ash and they weren't always the bigger pyre. It had been the coal which had been used which was of very poor quality even though it had been sold for a high price. Two months later after their pyre had been lit he could walk down and read the (ear) tags on the cattle. That's how bad it was.*

In April 2001 there were plans to build a huge 'metapyre' outside Longtown which would burn for weeks in order to dispose of the huge backlog of culled stock awaiting 'disposal'. This prompted an angry packed public meeting in the town, which was covered by the local newspaper:

People power stops the pyre: Longtown fury halts huge fire that would have burned all summer
(Carlisle News and Star, 12th April 2001)

Burning as a means of disposal was abandoned in the UK after this meeting, resulting in the use of large scale burial and incineration at rendering plants as the remaining options. The on-farm pyres also evoked responses on different levels at different times. In a very particular example of cultural clash, a young farmer wrote in her diary about attending the touring art exhibition 'Love Labour Loss' which came to Carlisle in spring 2002. This show included a Damien Hirst exhibit of a calf in formaldehyde:

Went to see art exhibition at Tullie House re livestock. Good exhibition except the calf! Very inappropriate to bring it to Carlisle after FMD. Reminded me of the calves we used to see in their mother's bodies when their abdomens split open on the pyre.

So particular combinations of people, minerals and machinery, which in a different space or time would be emotionally benign, become a trigger for feelings of dread and distress:

If I see a man, in a white suit, especially if it's got a hood up, I get a cold shiver down, my hair stands on, on end on the back of my neck. I saw too many, standing outside too many farms, you just knew it was probably not a good reason.

So meanings of particular materials shift about, becoming loosened or tightened in particular ways. In contrast with straw or farm vehicles, guns, bolts[59] and ammunition, while not everyday objects, *became* familiar during the culls and became everyday objects for slaughter and disposal teams. Every farm has occasional deadstock, so carcasses, while a familiar sight in ones or twos, are deeply shocking in large quantities; yet piles of carcasses both on the farm and in the media became a commonplace. As we explored in Chapter 3, death took place, *in the wrong place*. It happened out of place, time and scale. As Mary Douglas notes, materials which are 'out of place' still retain some identity because they can be traced back to their origins. With FMD this frequently involved piles of carcasses, caught up in the processes of dissolving, and rotting which would eventually render them unrecognizable. They were out of place. Lying in piles for collection, their disposal remained unfinished. 'This is the stage at which they are dangerous...their half-identity still clings to them and the clarity of the scene in which they obtrude is impaired by their presence' (Douglas, 1966:160).

A farmer told us about the horror of dealing with ewe carcasses, decomposing yet still recognizable:

The slaughterers, they started on the Wednesday morning and I went into the shed for summat five o'clock at night and you couldn't... you know the smell, they don't half go off quickly, when they're sick.

Similarly, the process of decomposition invaded sense of *place* on the farm:

They were left and of course they'd swollen up and then when they came to get them out they couldn't get them out. So they had to use you know loops round their feet, drag them out and then of course they also, because they'd blown up so much with gas then they had to be cut to release the gases and things like that..., but then at some point in the evening she'd had to go to get logs for the fire, she went out of her back door, and recoiled because this was a job that was supposed to be done finished.

The processes of disposal were not straightforward in this new context for people who had worked among stock in normal circumstances, and it was felt this was very poorly understood by people from 'town':

The town people were saying things like 'Well these farmers' lads can cope with it'. But these farmers' lads have been brought up looking after animals not destroying them and chopping them up so that they can be buried and burned.

The production of milk became highly problematic; while it became the only source of income for dairy farmers who could not move or sell

Figure 4.7 Farm cull

their stock due to the rural lockdown, yet the tankers which visited farms to enable that income, themselves became agents of possible infection and so of risk. Cattle feed was also associated with risk, an urgent necessity because of movement restrictions, but also loss of income, as a farm supplies manager explained:

> Until the end of August we didn't suffer too much, we've sold a wide variety of things, you know feed, fertiliser, seeds. So there's always something in season, during the summer, but once September came, you know the effect on our business was quite dramatic then, we were running about 50% of what our normal monthly turnover would be, so it looks as if we'll not make any profit at all.

On 'standing' farms (that is those who had managed to keep their animals) wool, usually managed through clipping and parcelled away, became scattered around fields as unclipped sheep gradually shed fleece. On culled farms, buildings, sheds and barns became denatured objects of punitive potential. Especially at the start of the epidemic buildings were required to be cleaned and disinfected to a rigorous standard, not always straightforward in large, old, stone-built byres and sheds, and inspections of the finished work were meticulous:

> It was monotonous. Because I mean after they got it all washed out, like all the cracks in, we'd to plaster all the cracks up and every little bit. Because there was one or two of the field officers you know the senior men, used to, we got them round once and he was 'gan about with a toothpick like and if he found the least bit hole, it would got, you know got a bit of muck out on it, he says he'll fail it.

Sources of information moved from the benign to the contested, became objects of confrontation: maps were drawn and redrawn placing new and sinister boundaries where no boundaries had previously existed. Familiar maps were overlaid with outbreak maps or risk maps or exclusion zones. Lists of farms began to read like lists of casualties in war. The act of listing, naming of casualties was also highly controversial and sometimes mistaken. 'Official' FMD and biosecurity letters, booklets, 'educational' videos were produced, to legitimize the altered significance of materials and practices. 'Unofficial' materials such as news reports, radio phone-ins, email, telephone contact, rumour grew in importance, as demonstrated by the size of radio audiences, use of helplines and extent of personal phone bills.

> I have never seen my husband get so mad ever as I have seen him this year, just because people would not listen to what you said. I mean even into October you would get a chap ringing up from Carlisle: – 'We want to come and advise you on bio security...', I mean having lived with it all year it just seems so pointless. 'Oh well, I have to ring you and ask you because I have been told to do it.'

After the majority of farms in a particular area (such as North Cumbria) had been 'taken out' in popular terms, or 'depopulated of livestock' as the official government letters told their astonished recipients (Figure 4.8), a different kind of *time* started to prevail. Gone were the daily rhythms of work, the sounds of livestock, the urgency of repairs or feeding routines. An infant school teacher explains how small children found the change distressing:

> They didn't like their farms, they didn't like their homes. One of the older ones, said to me 'I just don't like it.' I said, 'why?' 'When you go in the [cow] sheds, it's all spooky.'

80 *Animal Disease and Human Trauma*

Ministry of Agriculture, Fisheries and Food
State Veterinary Service
Animal Health Office
Hadrian House, Wavell Drive, Rosehill Industrial Estate, Carlisle CA1 2TB
Telephone: 01228-591999   Fax: 01228-591900

Mr   Date: 14 May 2001
                                                   Our reference: DC

Cumbria

To: All IP and DC Premises

Dear Mr

Foot and Mouth Disease Order 1983 - Cleansing and Disinfection

Because of Foot and Mouth disease your farm has been depopulated of livestock. In the near future you will be visited by an officer of the Ministry who will be able to advise you of the necessary procedures for cleansing and disinfection and how these will apply to your premises.

Figure 4.8   MAFF letter to infected premises

Figure 4.9   Altered sense of place

Recreating the social

People tried to recover what they had lost in a number of specific ways. As Harada (2000:210) highlights in relation to the Japanese earthquake, 'There is one sense in which materials have the same weight and size anywhere on the earth. But there is another sense in which they achieve their significance relationally in relation to other materials.' A poignant example of this can be found in the diary of a craft shop owner when he noted that:

> We then started getting people in who had lost their animals, had had compensation and were then wanting to buy something to sort of remind them of the animals they'd lost. So we would have like a Texel tup and Texel ewe and lamb, and people would be buying these things [limited edition figurines of farm stock] *to remind them of stock they'd lost. Sometimes it might be somebody who'd had perhaps several hundred and other situations, it might have been a young person who'd had perhaps one tup and half a dozen ewes.*

Others felt the need to record what was lost:

> *I said we should have video'd some of what's gone on here because people will never believe it or whatever, and what the cows were like, and what have you. And (Tony) just said 'where's your camera?', and just took my camera and went away outside and, I can remember him going out with the camera 'cos he's just not that sort of lad, he's a really sort of quiet, country lad, and I just thought God, you must be getting, you must be getting screwed up about this too.*

Art projects sprang up, site-specific sculptures of cows, bulls and sheep were made out of wire and placed in fields where 'real' cows should be. As a parent noted in her diary:

> *This week ended with an art exhibition in the village school. This was of the children's excellent work, guided by their art co-ordinator. She has helped them work through F&M with their art works. The school was full of villagers coming together for this joyful occasion and it makes me feel very emotional just to write about it.*

Famously in the tiny North Cumbrian village of Milburn, schoolchildren made a wire, clay and straw sheep and lamb which stood for a year

Figure 4.10 Milburn School 'sheep'

on the village green (Figure 4.10). In a further transformation of the everyday, the school then won an Artworks Award from the Tate Modern and a number of pupils at the school travelled to London to collect their award. When the models were finally removed (after vandalism) there was a sense of sadness:

> *Yesterday the 'sheep structures' from the green were removed for a final time. I chatted to the art leader who inspired the creations. She was very sad, although she said they were never meant to last, they were a great symbol, particularly for her, about F&M. The child who damaged them beyond repair was not involved with the making, had no idea of their meaning, and does not attend the school. Although some children had been seen to dump coats on the sheep, quickly reminded by people like yours truly, none have damaged them. They have been very proud as we all are. Many visitors have stopped to watch, examine and read about the sheep. They were taken away unceremoniously on the back of a coal lorry.*

Attempts to recover what had been lost were also connecting with an older sense of time, one relating to seasonal rather than clock time. These ushered in a post disease era in which people began to remember the events and seasonal ceremonies which had been banished from the local calendar. A year had been lost.

> *There was a ploughing match we'd went to and I thought, 'God this is the first time I'd seen farmers', there'd be no shows nothing to go to, no meetings, nothing and like I say, it weren't till September that we'd just realised how bad it had been, when there were all these fires just sitting on the roadside and in the middle of villages and this sort of thing.*

Here as documented so closely by Harada, 'things' and how people use them play a critical role in recovering 'the social'. Harada tells how 100 earthquake victims who had lost their homes were housed in a public shelter in a school building for up to seven months, sleeping for much of this time in a gymnasium. He noticed one day that families had taken up plastic insulating boards donated by a firm of building materials suppliers, to construct small 'houses' (they could still see over the top of the boards) held together with parcel tape, in which they arranged their few belongings and created a sense of 'home' within. Instead of using these boards to lie on as had been expected, they used the boards to create a personal sense of place from out of the disembodied or aggregated space of the gymnasium. Just like the elderly woman housed outside the disaster zone tries to recover her household dishes from the rubble, so we also see how the FMD community tried to piece together past and present, to reshape the present in familiar ways. Her 'old' house is to be bulldozed and she makes heroic trips back to the rubble, sometimes on her own to recover something of what the family knew before the quake. She wanted to recover some form of continuity in her life and that of her family.

'So materials created and represented distance and social miscommunication. That is they created difference. But at the same time materials sometimes represented family continuity – or the continuity of an individual's life, by being carried from the "inside" to a safe place on the "outside". Thus it seems…that a small china dish was able to represent the past decades; though the house or other belongings no longer existed' (Harada, 2000).

However, the recreation of the 'social' in an individual's life, only makes sense in a social context, as the emergence from the year long rural disaster of 2001 demonstrated, in 2002 respondents recorded in their diaries

> *The children can now play in the fields again. The first time they went in was when it snowed. They had great fun building snowmen, igloos and doing snow angels. It's hard to believe that it's a year since they had to stop going in.*
>
> *We couldn't have any village sports and any of the usual things that went on in the village because most of the people on the village hall committee are farmers and farmers wives. Now we are having meetings again.*
>
> *This week started with a very enjoyable Mayday parade in Penrith. The children had prepared their costumes and practised dance and music. Many people lined the streets and the children enjoyed parading for them. They particularly enjoyed banging heavy drums. It seemed a ridding of last year's bad experiences.*
>
> *Met a flock of sheep on the roads on Monday – which slowed me down but it was nice. My attitude to moving livestock has changed since FMD, previously these occasions were seen as a nuisance. Now I stop and appreciate seeing them.*

Personal experience and collective experience are intertwined. It is, we would argue, only through the personal accounts of disasters that their social and collective impact can be understood. This is why it can seem offensive, as when the 2007 outbreak occurred, to quantify and aggregate the 2001 disaster in this way:

Government must act fast to avoid new 2001

> THE 2001 foot and mouth epidemic is thought to have cost the country more than £8 billion, of which almost £3 billion was funded by the taxpayer. Up to 10 million animals were slaughtered, plunging thousands of farming families into financial ruin as the disease gripped the nation. Tourism was also severely hit as footpaths across the countryside were closed for months, with the cost to the trade put as high as pounds 3.2 billion.
>
> (The Daily Telegraph, August 4, 2007)

In disasters it is therefore impossible to overestimate the importance of 'things', whether this is the condition and presentation of personal effects following mass fatalities (Eyre & Payne, 2006), or the iconic and re-traumatizing effects of particular sights and smells for farmers and residents who experienced the burning pyres during 2001 (Mort *et al.*, 2005).

Summary: why detail matters in disasters as much as the big picture

Elsewhere in this book we have detailed the emergence of a loss of trust in authority which sprang from the perception of agencies engaged in the 2001 epidemic that local knowledge was unimportant and irrelevant at best, and obstructive and dangerous at worst. The wisdom prevails that disaster or emergency planning 'experts', who after all have no vested interest, know best how to manage a crisis. While the differences in nature and specificities between disasters, such as say the flooding of New Orleans following Hurricane Katrina and the 2001 FMD epidemic, are of course compelling, this marginalization of the 'local', of local knowledge is commonly found. The local is founded on proximal, *partial (*as in connected) experience, and stands in contrast with the distal and *impartial* (as in disconnected*)*, or 'objective' perspective. This marginalization of the local is made visible for example in the ways that particular details or materials come to be valued in disasters: the tangible and concrete ways those details are understood. As a 70 year old survivor stripped of everything by the Buffalo Creek flood said:

> I've often thought some of this stuff could have been avoided if someone would have come around and said, 'Here's a blanket and here's a dress for your wife or here's a sandwich. Could I give you a cup of coffee?' But they never showed up. Nobody showed up to give us a place to stay....The Pittston Company never offered me a pair of pants to put on, no shirt....That's what gets you kind of riled up (Erikson, 1976:181).

Why for example was it so poorly understood that the reason many residents would not leave their homes in New Orleans is because they had been told they had to abandon their pets in the evacuation? Carin Wittnich and her colleagues set up an 'owner finder station' during the New Orleans floods, recognizing that the possibility of human victims being reunited with their pets was deeply important for the recovery of both (Wittnich & Belanger, 2008).

During FMD There was anger that local knowledge and expertise was ignored and untapped:

> *Things need to be worked on a more local scale, you know, the people, trying to get an answer out of DEFRA Carlisle, 'oh we have to refer to Page Street in London' and this Page Street in London hadn't a bloody clue like, they just hadn't a clue like. The rules, they were just making up as they went a long like, and they'd no idea like.*

It is now part of FMD folklore that in the wake of the disaster, an official asked a farmer over the phone: 'What sex is your bull?' And that another requested that 'David Brown' be added to the list of farm residents approved for moving stock.[60] This notion of distal control seemed to be reinforced by the inquiry reports that came out in 2002:

> *Read some of the findings on FMD inquiry and how Cumbria handled it. It seems the same old findings, too much bureaucracy – too far away from the actual location of the outbreak, i.e. London insisting on all decision making.*

There is something to be said here about the material practices of empathy in dealing with disasters, and later in the book we discuss the need for 'therapeutic spaces' to be made available for people who have experienced a traumatic event. What <u>has</u> to be done is one thing, doing it with some imagination about its effects, its meaning for those on the ground is another. The ability to recognize the new identities and resonances taken on by objects and places in disasters is a part of this. Why for example, was the National Coal Board unable to understand that the people of Aberfan needed funds to build a new place of worship after the town's chapel had been used to house the bodies of their children recovered from the school engulfed by coal tip landslip? The Coal Board called this request 'irrational'.

> *Up until then I had hoped that the chapel was a hospital, but as I went into Bethania people were coming out who'd been told that their children were gone. Until I went in I still had hope that they were just lost. When I went all the pews were covered with little blankets and under them lay the little children. They*

picked up the blankets and showed me every little girl until I came to [her] and said she was mine.

(Bereaved mother quoted in McLean & Johnes, 2000)

To understand this is to understand what is meant by being 'haunted' by something, the way in which a thing, a place, generates new energy and meaning (good or bad) for those who witnessed something glorious or terrible there. How could that chapel not resonate with grief, forever, for those who knew what had happened there? So drawing on Haraway's cogent argument (1988), without situated knowledge which is local and *partial*, in the sense that it has a perspective based on experience, authorities cannot 'see' how to act, and this is one of the hardest lessons in the wake of disasters where the actions of recovery agencies have the potential to bring relief, but also to retraumatize and make things so much worse.

Here we are thinking both of what has been learned about of the role of agencies after the Buffalo Creek and Aberfan disasters, but ironically also in contemporary (Western) settings where 'disaster management', underpinned by 'emergency planning' occupies so many policymakers, guidelines, committees, study programmes, field exercises. Yet the power of the local, and the web of detailed associations *in* the local, are integral to the implementation of any strategy of (flood or disease) containment, disaster management or recovery. These associations are what give action meaning, make action possible. These associations of people, materials and places are also what give rise to expressions of solidarity to be found in almost all disaster sites, often against a prevailing centrally driven culture of individualism and self-reliance. For what was proved along the Gulf coast is that the lives of citizens are enhanced by, and indeed inseparable from, the construction of collectivities (consisting of humans and non-humans) that increase people's capacity to act, not by further dismantling of the institutions and infrastructure that enable people to live (Braun & McCarthy, 2005).

[FMD] *showed the inter-dependence of the area with agriculture... I suppose it's a bit like a coal mining community isn't it? You know, if coal mining goes down, you know, pit closes then the community is severely affected. Well if farming goes through a bad period, then the whole community is affected, you know the whole business community is affected. And obviously we've thought of this before Foot & Mouth you know when farming is not doing so well, well then that will be reflected in our business because many of our customers are involved in farming. So I think it highlights that, it highlights the inter-dependence of communities and businesses, and people. In many respects I think it brings people closer together. But then later on I think it will have driven some farmers apart.*

5
Trauma and Traumatic Experience

Introduction

Our study of FMD and other disasters has led us to consider how trauma, and in particular, the relationship between traumatic experience and the diagnosis of Post Traumatic Stress Disorder (PTSD) has come to be understood in such a context. Traumatic stress is perhaps better represented as the 'normal' reactions of those people exposed to an abnormal disaster event (Yehuda *et al.*, 1998; Alexander & Wells, 1991). As Secor-Turner & O'Boyle (2006) indicate, most people who experience a traumatic event will experience feelings of fear, sorrow, uncertainty, and anger; these feelings are usually self-limiting and last three to six months. Our frequent interactions with those affected by the FMD disaster have led us to develop a situated definition of trauma. In 2001 traumatic experience was widespread and both acute and chronic, and people have reported feelings of shock, depression, including thoughts of suicide; loss of concentration and interest and recurrent thoughts and flashbacks. However while some of these feelings and affective disruptions remain, and perceptions of normality might be changed in the long term, we want to argue, along with Erikson, that this is what people who have survived disaster share, but it does not mean that they are sick. Survivors may be marked (stigmatized) by what has happened, but what helps is to find ways to understand and connect with this traumatic experience, rather than to separate it from wider society, to classify this as abnormal or 'other'.

In a review of the trauma response literature, Secor-Turner & O'Boyle (2006) state that the way in which those involved experience a traumatic event depends on the nature of the stressors and the mediators associated with the event. The nature of potential stressors includes risk to life, exposure to death and dying, physical injury, duration of the event, loss, and degree of terror. Mediators (factors that moderate an individual's experience of traumatic stress) include anticipation or worrying about possible events, assessment of the threat, perceived sense of control, role constraint, and prior disaster experience. The perception of the origin or cause of the event, such as human-made versus natural, and the potential for stigmatization of victims also serve as mediators as to how the stressful event will be interpreted by individuals.

Bonanno (2004) argues that most people are exposed to at least one violent or life-threatening situation during the course of their lives, and whilst people cope with these potentially disturbing events in different ways, most are able to endure potentially traumatic events. Indeed, resilience in the face of traumatic events is not rare but relatively common, and represents healthy adjustment rather than pathology. Significantly, Bonanno (2004:20) differentiates between resilience (the ability to maintain a stable equilibrium) and the concept of recovery (where normal functioning gives way to threshold or sub-threshold psychopathology, usually for a period of at least several months, and then gradually returns to pre-event levels), arguing that resilience represents a distinct trajectory and is more common than has been understood.

Jones (1995:509) argues the concept of PTSD[61] fails to 'embrace the complexity of the experiences of suffering and loss in [disaster] situations.' We would broadly support this perspective (and the work of others such as Bracken *et al.*, 1995; Summerfield, 2001) and argue for a more nuanced, situated understanding of traumatic experience in disasters, which may share characteristics of clinically defined trauma, but crucially does not attempt to systematize and pathologize traumatic experiences (with related dangers of disempowerment and victimhood). As Summerfield (2001:95) points out, PTSD is socially constructed:

> A central assumption behind psychiatric diagnoses is that a disease has an objective existence in the world, whether discovered or not, and exists independently of the gaze of psychiatrists or anyone else....However, the story of post-traumatic stress disorder is a telling example of the role of society and politics in the process of invention rather than discovery.

Many people who experienced FMD in 2001 found personal and social resources which protected them against developing illness as a result of exposure to the repeated horrors of the FMD crisis. McFarlane (1988,

2000) observes that there does not appear to be a simple relationship between distress and psychiatric illness. Thus whilst Ursano et al. (1999) argue that disaster workers are at risk of both acute and chronic PTSD (where the most difficult aspect is exposure to violent death and dead bodies), other work notes the 'unexpected resilience' of rescue workers and body handlers working in extremely stressful conditions following the Oklahoma City (USA) bombing[62,63] (Tucker et al., 2002; North et al., 2002). A meta-analysis of five separate disaster events carried out by Cardeña et al. (1994) identified that the numbers of people displaying post disaster PTSD-like symptoms were very similar and were apparent in only a minority of respondents. Similarly, Secor-Turner & O'Boyle (2006) report that whilst the majority of people will recover without any long term problems, some who experience disasters (including disaster workers, which we will discuss in Chapter 7) will suffer clinically significant reactions.

Other studies have highlighted how the timescale of the disaster is important, not least because of the risks from continued exposure to stress and trauma. As Robinson & Mitchell (1993) indicate, experiencing cumulative stress is potentially more damaging than stress from a single incident. With FMD, the length of the disaster, combined with factors such as the clean/dirty regime imposed on workers, sense of loss of control, and transience of (some) FMD workers, weakened the ability of 'teams' to 'pull together.'

Anniversaries can be particularly difficult times for trauma-exposed individuals, and around a third have a likelihood of experiencing or exhibiting significant distress on the anniversary of their traumatic events (Morgan et al., 1999) In 2002, ostensibly the year which followed the 'FMD year' many rural people were forced to revisit the events of the FMD year through anniversaries of culls, of financial disaster or of their reluctant participation in the mass killing. Processing painful and distressing images and some degree of 're-living' the events of the past year may be part of the recovery process if that experience can be shared or acknowledged in some way. A slaughterman speaks of his feelings one year after being involved in the mass culls and records in his diary:

> It's a year since I went away killing. I feel a bit funny with myself today. It was our wedding anniversary on the 10th, but this sticks in my mind more.

A DEFRA worker writes of his relief at signs of renewal:

> It was good to see sheep back in the fields and even the first lambs were appearing again. The last lambs I saw last year were being given lethal

injections by a vet and we laid them out in rows to spray with disinfectant till they were taken for disposal.

With specific reference to the FMD crisis, we have argued that it is important to recognize the levels of distress experienced by farmers, their families, nearby residents and disaster frontline workers, highlighting the need for what counts as evidence in health research to go beyond the pathological and the statistical (Mort *et al.*, 2005). It should be recognised that many of the individual accounts of traumatic stress are accompanied by accounts of how individuals found a way to contextualize this, or gain some strength from the knowledge that many others locally were experiencing similar horrors. So while undoubtedly within the context of the FMD disaster, as something experienced individually and collectively, there are certainly features common to the clinical individualistic (e.g. DSM III) definition of PTSD, there are also features of *collective* damage and impact such as described by Erikson (1976) in his in-depth exploration of the after effects of the 1972 flood in the mining area of Buffalo Creek in Appalachia.

So the 'FMD year' represented both an *individual* and a *collective* trauma in Cumbria and other parts of the UK badly affected. Following Young (1995) we can see how trauma is a heavily situated, contextual concept and our working definition for this book comes from what we learned from survivors in Cumbria. It also encompasses both the events and how those events were experienced by individuals and communities, and we refer not only to the distress, horror, and re-traumatization but also endurance and sources of support articulated by those who experienced the events of 2001. The trauma of losing generations of work in building up pedigree herds and flocks at a stroke; the uncontrollable effect on rural businesses threatened by the economic impacts of restricting livestock and human movement; the social impacts of closing auction marts, cancelling seasonal rural social events and isolating 'infected premises' (which means isolating human beings); the public health concerns of carcass disposal methods including landfill, mass burial and burning on pyres; all these may have profound health and social consequences. The human consequences of these disease control strategies have, however, been afforded little discussion. Yet as the Cumbria Inquiry (2002:57) notes, the environment and the lives of local rural communities were transformed: 'these events caused a huge level of trauma and distress for many in the population, and brought attendant concerns about short-term and long term-problems of health.' As both Chapter 3 and Chapter 4 have demonstrated, the experience of having 'life's work' destroyed was deeply traumatic for those on the receiving end.

A disaster realized

In Chapter 2 we drew on a range of resources to give a broadly chronological account of 'a disaster unfolding'. Here we connect narrative and local experiences of the social, cultural and economic consequences of FMD and in so doing, consider 'a disaster realized'. We consider themes of closure and disruption, work, disjuncture and distress. People's recollections of the 2001 FMD disaster and ongoing consequences, some 18 months to two years after the initial shock, have common elements. There are first memories of the FMD outbreak (beginnings); initial disbelief leading to the disruption of *normal* routines, which for many led to feelings of discord, everyday rhythms and familiar places, becoming disharmonious; in turn fostering a sense of separation, disjuncture, feelings of being disconnected and finally in some cases, distress arising from an overwhelming sense of loss, hopelessness and despair:

> *It was the talk of everybody that came in* [to the shop], *and the fear within people, you know, it's too near, it was a funny sort of atmosphere.*

A farmer notes:

> *We were just sort of hanging on, watching the news and who was getting it. From the 12th to the 25th of March, they were just going down round us like nine pins like. You know, it moved on to the next door farm, next door and next door, by the 25th March we were the only one left, all our neighbours had gone, and we wondered why we hadn't got it* [then], *it started up again and started coming down towards the end of April beginning of May, and cleaning one or two up that it had forgotten the first time. And, it swept us past again.*

There is also recovery, a sense of experiencing life before FMD, life during FMD and life after FMD. The 'before' and 'after' are not the same, they are forever punctuated by the FMD disaster phase in between. A community nurse recalls:

> *Going out to Penrith last night my eyes automatically duck down where the pyre smoke used to be. It's not a conscious thing, it just was present*

94 *Animal Disease and Human Trauma*

for so long and was so depressing, so synonymous with the despair of that time that I look towards it, just as I always check what the hills are like each day. I suppose its become a part of my subconscious landscape. At least if there's no smoke or smell then the worst must be behind us.

Being shut down, cut off

On Saturday 24th February, the banner front page headline of the North Cumbria newspaper Carlisle News & Star read 'Keep Out!' The corresponding article discussed the closure of marts in Carlisle, Penrith,

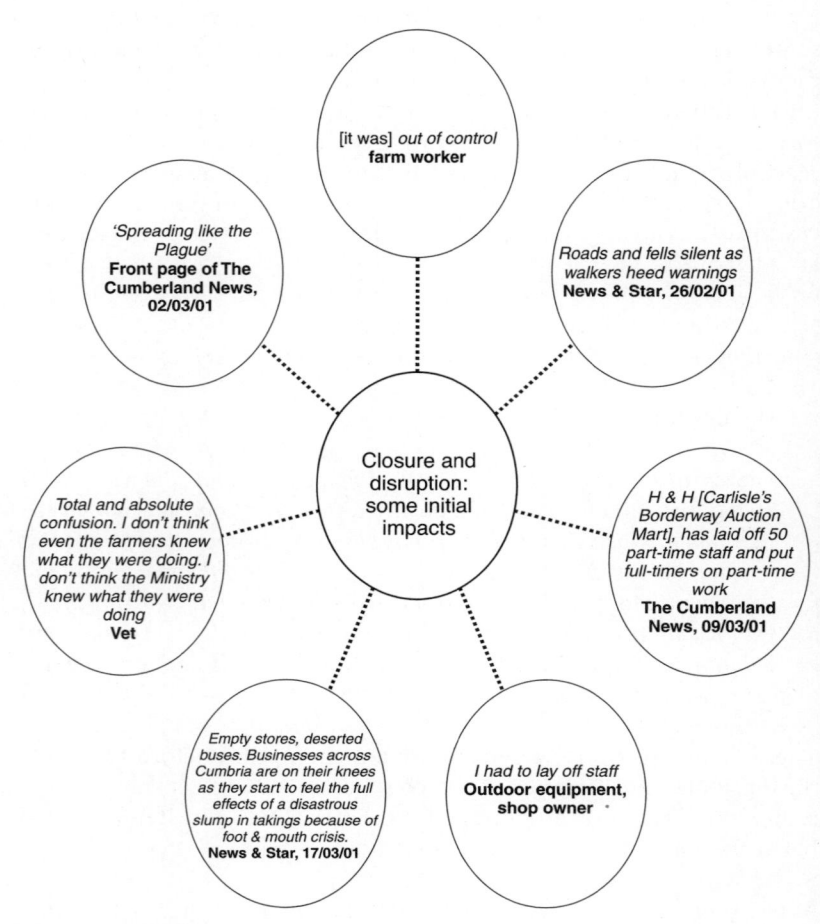

Closure and disruption: some initial impacts

- [it was] *out of control* **farm worker**
- *Roads and fells silent as walkers heed warnings* **News & Star, 26/02/01**
- *H & H [Carlisle's Borderway Auction Mart], has laid off 50 part-time staff and put full-timers on part-time work* **The Cumberland News, 09/03/01**
- *I had to lay off staff* **Outdoor equipment, shop owner**
- *Empty stores, deserted buses. Businesses across Cumbria are on their knees as they start to feel the full effects of a disastrous slump in takings because of foot & mouth crisis.* **News & Star, 17/03/01**
- *Total and absolute confusion. I don't think even the farmers knew what they were doing. I don't think the Ministry knew what they were doing* **Vet**
- 'Spreading like the Plague' **Front page of The Cumberland News, 02/03/01**

Figure 5.1 'Shutdown'

Cockermouth and Longtown, which had been shut the previous evening along with a nationwide livestock movement ban in an attempt to slow the spread of the disease and a plea from the Lake District National Park Authority officials to the general public, to 'stay away from the Lake District'. As the Cumbria Inquiry Report (2002:19) indicates, the epidemic led to a countrywide ban on animal movement and to widespread restrictions on public access to the countryside. Coupled with media images of animal slaughter and disposal, the restrictions on access and freedom of movement brought many businesses in affected areas almost to a halt – tourism and all types of outdoor recreation were particularly badly affected. Figure 5.1 indicates some reactions to these initial stages of the disaster from both respondents and the media.

In March, the front page of the Cumberland News (09/03/01), spoke of 'Cumbria's Killing Fields' and how 'MAFF can't cope with numbers', animals were being 'left to rot on farms.' The disaster began to disrupt every facet of daily life. Familiar places were described as: *unreal, bizarre, crazy, surreal*. A slaughterman noted that *when we first started killing the stock as soon as they lit the fires everything went silent. There wasn't a bird singing in the sky, there was nothing and it was just an eerie feeling*. Other people talked of how there *was nobody in the streets*. Some changes were still palpable in 2002: *Things still don't look or feel 'normal. There are too many unpleasant associations in the fields and pyre sites; it still seems too recent*. Similarly, someone else notes that the: *Fields are very quiet in North Cumbria, a little sad feeling. It's amazing how activity i.e. animals in fields, affects you*.

Whilst the media focused on visitors who were denied access to footpaths, access restrictions also affected residents, unable to walk dogs, take 'time out' from the strain of living under siege: they were trapped. For those living near disposal sites their local roads were the conduit for the disposal of carcasses with constant streams of lorries and a feeling of being invaded, even violated by the 'mortuary production line of previously unimaginable proportions' which served to funnel 'the dead livestock to this small number of hastily selected sites...dictated by the operation of the slaughter policy, whose draconian form precluded any but the most cursory public consultation' (from a study of FMD in Tow Law, County Durham, Bush *et al.*, 2005:650). Yet this stood in contrast with other places which were unprecedently quiet, with footpaths closed, even walking on the roads in other areas made some feel guilty and they spoke of *ghost villages*. So for many, but in different ways, this landscape and these spaces are changed forever:

> *The sheep, cattle and ponies are no longer grazing the fell, so no longer meander into the village in the colder weather. We always knew when a cold snap was due because the fell ponies came down to the village. I guess that I miss that aspect of life. We hoped everything would get back to normal after FMD was over – but much of it has changed to a different 'normal', a less inclusive one.*

A person living near a disposal site writes of how this change in her landscape has brought about irretrievable loss:

> *It was a lovely place to go walking and to relax by getting away from everything.* [since the site itself was previously a place of recreation] *It used to be lovely and quiet and was also so nearby. We do actually quite miss it sometimes as there is nowhere else like that round here now.*

For many, there was a change in their 'sense of place', which is perhaps most simply considered as an overarching concept describing relationships between human beings and spatial settings (Shamai, 1991). It is also about situated social dynamics and is multi-dimensional, holding different meanings for different social groups. This relationship to place is an important source of individual and community identity and provides a profound centre of human existence to which people have deep emotional and psychological ties (Teo & Huang, 1996). Egoz *et al.* (2001) state that sense of place is heightened when change threatens existing landscapes, and for those living in the worst affected areas their immediate surroundings became hostile, bleak, beleaguered, and (from their perspective) unhealthy.

Personal relationships

Family life was also disrupted. Family members remained isolated and separated for months because of fear of spreading/acquiring infection or through working away from home. However there were also many accounts where respondents expressed pleasure in spending time with their children, with the 'hope for the future' that children afforded them. There were children who did not attend school for weeks, sometimes months, missing both education and social interaction resulting in longer term problems:

> *Spoke to a friend, who is having a really tough time with her* [teenage] *daughter, she won't go to school. They think its delayed shock from last*

year's FMD. She hoped she would go back in September but she didn't, it is so hard for them.

Some parents educated their children at home using materials sent by the school. There are references to 'lost time' and guilt of family members who feel they missed seeing or paying sufficient attention to their elderly relatives. People spoke of the guilt they felt at the time not spent either with young children or with elderly relatives. A business woman told us how guilty she felt when her mother died in the winter of 2001, she believed that her own exhaustion and worry over financial difficulties had communicated itself to her mother, and that she had not had the energy to give her mother the time and attention she deserved.

> *I suppose a bit of me thinks probably if we hadn't had foot and mouth (close to the village) my mum...probably wouldn't have died last year. I think she was very affected by the way we all were and she could see that we were all very, very down. I think it probably affected her...I feel that it probably made a difference. I think the fact that we were very depressed about things and obviously worn out must have made a difference to her. She would have known that helping her was one more effort which is a shame.*

Some felt distracted and unable to concentrate on domestic work, others reported how they had 'discovered' gardening as a therapeutic activity. There was also expressed guilt surrounding lost family time and this was poignantly captured in accounts of 'FMD babies'. These were babies born during the crisis and for some people, whose home lives were disrupted by long working hours, there was a sense of losing irrevocably a significant period of their children's lives:

> *G's first birthday, feel very glad I bought the video camera because after the last year on FMD, I feel as if I've missed out on him growing in his first year of life – which makes me feel sad.*

People also spoke of weddings postponed, birthday and anniversary celebrations cancelled, and funerals held in private with a minimum of family present. For example, a farmer had kept away from a family wedding in another county:

> *We missed three weddings this summer because of foot and mouth. Just not gone. One of them, when my nephew got married in July and his*

reception was (in neighbouring county) and people were ringing him and saying if we were coming to the wedding they weren't coming.

In another instance, a funeral was held in private in order to *to spare people making a decision whether to go or not*. This disruption to the usual funeral arrangements was especially hard for some, as in communities where families have lived side by side for generations and are woven into the fabric of business and social life, it is traditional for funerals to be large occasions both for paying of respect to the dead and a communal acknowledgment and support to the living. The fear of travelling, of feeling peripheral, marginalized and of mixing with others is hard to recall now, yet it not only stopped people shopping and socializing, it kept families apart too.

In a way similar to the people of Aberfan in the years after that disaster, for many Cumbrians there was a sense of being 'too far away to matter'. As McLean & Johnes (2000:224) put it: 'The failures to prioritise the interests of Aberfan were not because it was Welsh or working class, but because the community was peripheral. Its cultural and social values were so remote from metropolitan ones that neither understood the other.'

> *I tend to think that the national press regarded Cumbria as a forgotten area. I mean it was bad in Devon, but most reports were centred on Devon. It was bad there, but here it was much worse. It was a forgotten area. There were reports in the national news but not as much as Devon. People in, in cities, I, I think they're pretty clueless really what goes on in, in the countryside, they're so far removed from the way you live.*

There was a sense that distance led to a tenuous grip on events and a failure to grasp the enormity of what was happening on the ground:

> *But the government just didn't care...It was just like we were in the middle of nowhere, London was miles away and they, all the noises that was coming from Carlisle were 'Oh we've got it under control'. They didn't have eh? It was just gan [going] wrong wrong and wrong.*

A vet writes:

> *The people in London didn't, or they didn't appreciate, or they didn't want to know, what ever way it was, they let us down. The guys on the ground,*

the vets on the ground, knew what the problem was and how to address it. And that's why I feel so badly let down.

Routines and regular events

Previously 'taken for granted' cultural activities, in particular, social events, routines and festivals were cancelled. Such community-based festivals are important as they often grow over time to reflect the values, interests and aspirations of the community. 'A sense of community and place are linked to such events...they reflect the community's sense of itself and its place' (Derret, 2003). These ranged from large scale: for example horse trials; county agricultural shows to smaller events: for example indoor bowls in a village hall and outdoor religious services. Indeed there were repeated, frequent references to the cancellation of familiar events during 2001, often when respondents were writing about such events being resumed the following year. The loss of social contact and recreation is very significant for individuals and communities and is known to affect well-being and general health. There was also a 'collateral' loss of charitable/local income from cancelled events.

Community residents tried to continue to offer both practical and emotional support despite the voluntary and enforced social isolation brought about by both the bio-security management of FMD and the related 'cancellation' of community life. Examples of neighbourly solidarity were characteristic of the FMD crisis:

> *One of my neighbours rung me up one day and she said there is a queer big box of groceries on your front doorstep. Because I had never been out of the house I didn't know that they were there even. We received all sorts of things that were just left at the doorstep, groceries, flowers, it was overwhelming.*

A culled farming family told us: *we have received over 100 letters and cards.* A rural vicar, whose normal practice was to visit and be with those in trouble, found that *the only way you could be with the farming community was on the end of the phone.*

When social events were held, they sometimes coincided and interrelated with the traumatic events of FMD, and the vicar relates the following chilling story:

> *Some parishioners were having their golden wedding and they were having a service of blessing [...] and then they were having a do at the* (village) *pub. So the night before [...] I had been to lock the church up and I popped in*

> the pub for a pint. It was [Aug 2001] about quarter past nine, and there was a vet and I thought, 'hello trouble'. The vet didn't come at quarter past nine unless there was trouble. So I came home and said to [my wife] I think [...] has gone down, there was a vet there. And we got this golden wedding and people coming from all over the country and all over the world. [...] We had the blessing and I asked people to leave their cars by the church and walk up to the village, it was only 500 yards, wherever possible and they were very good about that. And there was this celebration going on and I just couldn't be part of it, we couldn't be part of it somehow. And we went out, the sun shone and the guns went off, you could smell the blood in the air and the wagons came and then the word came that [neighbouring farm] had lost them all, killed all as contiguous and his wife had died at home. They had culled all the lambs and she had died during the night. Unbelievable, it was just unbelievable and we just couldn't celebrate. Villagers came and stood on the cross by the post office and we just stood, there weren't any words. [...] Dreadful, dreadful day.

Ongoing local traditions were sources of comfort and offered a sense that some things would endure in the face of mass slaughter and loss:

> The one thing that survived FMD last year was the annual 'Dole Ceremony' at the Countess Pillar on the A66 at Brougham [...] as far as we can tell it has taken place on the same day at the same time every year since Lady Anne Clifford inaugurated it in memory of her mother in 1654. There are some things that even foot and mouth could not kill off.

Another person speaks of visiting her mother's grave on Mothering Sunday 2002 and how:

> It was pouring down with rain. Then on the way to Castle Sowerby church (where husband's Mother is buried) we passed two sites where carcasses were burned, the ash has been removed, the ground levelled and reseeded. The physical traces of what happened are easier to erase than the mental ones.

Working practices

There were many references to work patterns being disrupted. In Chapter 6 we will discuss the issues affecting frontline workers in disaster situations, but for every rural occupation work changes dramatically during an FMD outbreak, as has been clearly shown most recently by the 2007 crisis. However when that outbreak becomes an epidemic, work is transformed in unforeseen ways. For example, wagon drivers faced new imperatives when collecting milk or delivering feed and were under constant scrutiny as to bio-security. Farmers reported frustration and anger at perceived inefficient, bureaucratic and incomprehensible movement licencing arrangements. A vet notes how:

> *On the first day that movement licences came out we applied for 26 and received the explanatory notes from DEFRA on how to complete the paperwork four days later.*

Later he recalls talking to a farmer who

> *....lost no stock, but he was so depressed he started crying. The cows I had gone to see* [after FMD 2001] *'should have gone last year, then we wouldn't have had these problems'. He also has been unable to sell sheep so the farm was completely overstocked, overgrazed and had little silage left. I couldn't understand why he was still overstocking when he could have sold sheep and cattle (to other farmers) for restocking or for slaughter quite easily in the last few months. Maybe he just couldn't face the hassle of getting licences, or maybe he's just too far down to think straight.*

Biosecurity meant a particularly invasive level of surveillance and perceived lack of trust by the authorities, as well as licences being required for all animal movements, licences were issued to each member of the family on culled farms to come and go from their homes. Any machinery or equipment brought onto farms by contractors (to bale silage for instance) had to be licenced on and off with details of vehicle registration numbers and maps showing the location and identification numbers of the fields to be worked. The sense of cultural isolation was reinforced when farmers trying to obtain licences by telephone faced frustration when Trading Standards staff did not understand farming terminology. One farmer spoke of having to fax through a drawing of a 'creep feeder'[64] that he wanted to move to a field away from the main farm, because the person granting the licence could not grasp what it was.

At the end of October 2001, the busiest time of year for sheep breeders who would normally be buying and selling breeding stock, female sheep and rams,[65] the National Farmers Union contacted their members with a plea for patience.

> There is still a backlog with the issuing of sheep licences with about 21 days between application and receipt of the licence. DEFRA reported

today that they had received 1642 applications for sheep licences since the autumn movement regime started on 1st October and had 450 still waiting to be processed. They have been drafting in additional staff in the past few weeks to cope with the backlog but have had eight people leave in the last week because of abusive phone calls.'[66]

In the longer term, the restocking of farms brought animal health problems and continued stress and reminders for farmers as they became familiar with their new livestock. Indeed, the process of restocking has been a contradictory experience, bringing both renewal and sadness at the loss of often irreplaceable bloodlines.

> *Vera was talking about this* [restocking] *on Wednesday, saying how, one morning she was trying to bring in her new cows for milking. They went to the corner opposite the gate. It was sheeting with rain and as she struggled she looked across the valley at her neighbours getting their (un-culled) herd in. She says she thought 'You thought we were the lucky ones because we got paid out – you should try this.' This was all related with her usual good humour, but it's unusual for her to be anything other than completely stoical and accepting of circumstances.*

Farmers have spoken of a *loss of confidence* in handling the new cattle; an increase in calving difficulties because the *wrong bull* had been used and vets commenting on an increase in caesarean deliveries, other calving difficulties and animal health problems such as TB resulting from the import of new stock from other areas of the country (Convery et al., 2005). Others reported feeling less involved in their work, sometimes over-reacting to problems among livestock. A community nurse notes:

> *Saw a patient who was injured by a cow calving. It was new stock, having lost everything to FMD. It changed her. After two weeks in hospital with surgery, it will take her six months to physically fully recover. I don't think she'll ever have the same confidence, at present she doesn't want to calf again.*

In some cases there was a 'redistribution of employment', here a slaughterman made redundant from the abattoir due to FMD, joins an FMD slaughter team:

> *We went back the next day on the Thursday and we never done anything and he called us all intil like a meeting eh, and there was about maybe*

60, 65 of us. And he just says 'I'll have to pay you all off', so that was it. And then he phoned on the Tuesday night and he asked if I'd go back and work for him, going round farms, slaughtering.

An agricultural supply business tried to keep its staff on but they lost more than 300 clients through confirmed infection (one-third of Cumbrian cases) and the role of the business owner changed from supplier to confidante:

> But as the number of cases grew…we lost eventually about 90% of our customers, and it was obvious that it was going to take a long time to recover, so we then decided to make four people redundant, so we dropped down from 12 staff to eight, I went onto half pay….Frequent discussions during each working day (or evenings) with customers who lost stock. They all seem to want to confide in other people…this can be wearying as they all demand a level of attention and they are all different.

Income from tourism in the marginal areas was severely reduced and remained so throughout the whole of 2001. Small businesses reported securing bank loans to keep afloat during months of uncertainty.

> Our foot and mouth loan has another three years to go, […] It just goes out of the bank so you know that every month it doesn't matter how much you make, you're not making as much as you did before, and you're still paying back the money that you borrowed to get over the time when nobody came.

Some rural businesses in marginal locations suffered severe isolation and hardship. A craftsman, a sole trader who had just set up in new rented premises in January 2001 saw his hopes and optimism disappear with the visitors on whom his trade relied, and suffered severe isolation and hardship which went on long after the epidemic was over.

> ….on the back of having run through the quietness of last year it's too much sometimes, like for God sake's when is it going to end? It's as if there's a continuation of Foot & Mouth Disease, to some extent there is […] it's caused a huge slow down, for example, the footpath restrictions have been lifted but there still isn't anybody about. This is very much a walking area and the tea room here is reliant on walkers. […] The silence, the silence, no it gets sad sometimes when you're in here either in the big showroom here, tea room, a workshop with the machinery. The green

> outside, a nice little quiet hamlet and things on the shelf. I like interacting with people but then to return home to an isolated area to have my quiet solitude, that's me. And to be in here for a whole year every day to see nobody [...] you begin to hate the place, you begin to hate the very thing you love. (If I gave up it would be) *not because of the Foot & Mouth epidemic but because the lack of awareness by our Government to realise the human dilemma, the psychological effects that something like that can have on people's lives, aside of the finance, the economics, everything... people matter.*

However businesses in towns and villages in the central areas of Cumbria fared better. While hospitality providers on the *fringes* and in farm settings lost almost all their trade, the hotels and guesthouses in towns in the worst hit areas gained business from incoming vets, field officers and slaughter teams. DEFRA alone spent over £4m on staff accommodation.

Disjuncture and distress

Feelings of disjuncture, of being disconnected were described from within different situations and yet had very common elements, particularly a surreal quality, a sense of being somewhat removed, an unease that all was not as it should be. War imagery was often used:

> *By the 25th March we were the only one left, all our neighbours had gone, and we wondered why we hadn't got it. We didn't dare go out because everyone that did go out was getting foot and mouth.*

Later he describes going out on his first visit to a local event in September that year:

> *We came out as if we were coming out after a bombing raid, and you were just seeing what had been bombed.*

Disjuncture was also evident in the feeling of being disconnected from those 'outside' the disaster and the frustration of not being able to explain what it was like from the 'inside'. A rural vicar writes about struggling:

> *.... to explain what it was like to be here through last year but am still unable to articulate thoughts and feelings accurately. It is so hard to try to*

enable people who weren't here to get a real grasp of it.. The fear, the anger, the frustration.'

Accounts of the FMD crisis show distress, loss, feelings of despair, a shared sense of shock, horror and endurance, characteristic of other disasters. Distress linked to trauma also affected those not immediately involved in farming: a sole trader in a rural craft business saw his business 'wiped out' and as part of bio-security measures, was temporarily evicted from his farm based home; a camping and caravan business lost 90% business revenue and a year on, was struggling to remain viable; a health visitor who spoke of her fears that a family separation due to domestic violence, a stress related and incapacitating illness, redundancy and subsequent housing problems were indirectly related to the distress of FMD. As a rural vicar noted:

This has been a bereavement for individuals and communities. You have got to look at a minimum of two years to work through it.

Implicit from the diaries is a shared sense of shock, horror and endurance, emotions which are characteristic of other disasters. The FMD 'year' represented a *collective trauma* in Cumbria, a term developed by others (Erikson, 1976, 1991, 1994; North & Hong, 2000 & Sideris, 2003). We know that a wide range of livelihoods were threatened or destroyed; in farming generations of bloodlines and breeding stock were lost; family relationships were disrupted and damaged and many frontline workers were deeply traumatized by their experiences of working on slaughter or disposal, many extended farming families were cut off from each other for almost 12 months, many children lost months of schooling and relationships with neighbours were severely damaged by the imposition of clean/dirty regimes. In some cases, relationships with neighbours were damaged by the threat of infection being passed from farm to farm. Post FMD infection, others found it hard to go out at all, they felt contaminated:

And now he's feeling, his wife can't get him to go out anywhere because he feels dirty you know. And you know it's a hell of a thing like. I mean some people have gone out and doesn't seem to have bothered them. But others really conscious of having had it.

Importantly, for some the particular trauma of FMD is inextricably linked with processes of recovery. There are strikingly similar findings from a study by Bernadette Hood and Terence Seedsman (2004) of an outbreak of Ovine Johne's Disease[67] in rural Victoria, Australia. Culling was described by farmers as traumatic and emotionally shattering: 'We could hear the lambs bleating even after leaving the sheep yards, and we were no longer able to watch.' Hood and Seedsman write that farmers who were forced to kill their own flocks early in the programme were the most profoundly affected.

Experiencing the culls

Hall *et al.* (2004) argue that the likelihood of 'disenfranchized grieving' increases considerably in the event of livestock loss on farms in a disaster. If a decision to euthanize an animal has been involved in the loss, this may aggravate a sense of guilt, regret, or even failure. But the demand to cull healthy herds to contain an epidemic may be particularly devastating, as was the case with FMD in 2001. The culls were completely outwith ordinary experience and for many, the memory is enduring. Again, we refer to the study of Ovine Johne's Disease (OJD) in rural Victoria, Australia. where a farmer commented: 'I used to think of the killing fields. They were in Cambodia. I now know that they are here. I can't go back into that paddock.'

In the following extract, a farmer never refers directly in diary or interview to what took place on her farm during the cull. She refers instead to *that day,* shorthand which assumes an understanding of a wealth of meaning, and how as in times of war, sources of information become critical. She became distressed when she remembered:

> From [then] *onwards we were lucky if we had three, four hours of constant sleep, you were just lying there, you know just thinking of, about what had gone on really.[....] We got all our information from Radio Cumbria* [hourly bulletins] *and often you knew eight out of those nine people and you know it was hard to deal with.*

She describes the anniversary of the cull:

> *Anything we do this weekend will be better than last year. A couple of sleepless nights as the memories come back.'* A year later she says, '*I don't look forward to this week in the end of 27–28th April 2001, it has awful memories.'*

A farmer becomes upset in an interview by memories of his son's distress at losing his own stock: *And he's bred them...When we were fetching them in to slaughter like [...] you know, he was in tears...That was the worst.* He describes enduring the experience of the cull as something *you just have to block it all off as far as I was concerned* and the realities of decomposition, particularly *by the Saturday they were just all coming apart like...they were that heavy in lamb, if you just touched them like they just exploded.*

A year later his memories are still vivid, when every week for five weeks he recalls the events of the previous year in his diary: *this time last year the nightmare had begun... it was certainly the worst week of my farming life.*

Elsewhere in this book we have argued for recognition of local knowledge and expertise in disaster situations. During the Australian OJD outbreak, an approach was developed late in the epidemic to support farmers whose livestock was about to be culled. Following the diagnosis of the disease on a farm, the farmer's local vet would visit the family with an offer of help and ongoing support through the process of culling the herd. A second visit was made a day or so later, giving the farmer a chance to reflect upon the diagnosis and discuss (with a trusted figure) the management of the diagnosis on his/her farm (Hall et al., 2004). With due consideration of issues such as disease control and timeliness, something like this approach during the 2001 UK FMD epidemic would, we believe, have lessened the trauma of the culls for the farming community.

'Collateral' trauma

A teacher heard the stories of parents: *you had to listen ... it could be quite traumatic.* Her daughters' partners both found work on carcass disposal and they too told their stories: *One was handed a big long sharp piece of metal and told to run through a cattle shed and stab each one in the stomach as he went...he started having nightmares.*

A sole trader in a craft business located in a marginal area told how he was affected by a cull next to his work premises:

> MAFF came and knocked on the door. They sort of asked a few questions and I explained that I live on a farm ten miles away and they let me know that I couldn't return home there and weren't really able to offer alternatives...I had I don't know, £50 in my pocket and the clothes I was stood up in and nowhere to go basically, the first two nights the Saturday and Sunday nights, I slept in the car I found a lay-by away from, far away from here, I was just sort of driving aimlessly keeping warm most of the time. I tried a few helplines that MAFF had handed me and the

conversation with a lot of them was 'I'm sorry it's a sort of unusual circumstances, there's nothing we can do to help you'. The last one suggested I get in touch with the local authority which I did on the Monday morning and they housed me in a guest house. I was there for two weeks, that two week period was a nightmare.

He found ways of coping: *You almost hit a point where you're almost delirious and nothing matters. You don't care, nothing matters in the world and that was a comfort, that feeling was a comfort. You just switched off. And to a point I am still fairly well switched off.*

He later developed symptoms: *When I'm lying in bed I get palpitations. You're thinking the thoughts of the day, thoughts for tomorrow, what's going to happen next, am I still going to be in business, what if I don't, what am I going to do next? And then the panic attacks set in. I remember first time I saw the doctor he said, 'look there's lots of classic signs of anxiety and panic attacks, I'm going to prescribe you some more anti-depressants'.*

The way that many families were incarcerated for months, and the clean/dirty regime imposed on workers and many other rural citizens, also weakened the ability of communities to 'pull together'. But many accounts of traumatic stress are accompanied by accounts of how individuals found a way to contextualize this, or gain strength from the knowledge that many others locally were experiencing similar horrors. For example a vet told us that *it's something I shall look back on, not with fond memories, but definitely as an interesting part of my career, and at the moment I don't feel it's going to give me any long lasting adverse effects.*

An Environment Agency worker relates how: *you can only appreciate it through having experienced something similar... And I think for a long time I was looking for that kind of support, rather than going back into the office and, people coming to you, but without really having a clue of what it was like.*

Sharing stories

This collective telling, the shared recollections of the FMD epidemic (disruption, discord, disjuncture, distress), the lay ethnographies, are also specific to time and place and to local cultural understandings of such events. As such FMD narratives have resonance with community responses to other traumas. Sideris (2003) stresses the importance of connections between the individual and the social, and between local specificities and broader social structures in framing individual responses to trauma (in the context of the experiences of war survivors), and the work of Erikson (1994: 231–233) in particular highlights how disastrous events such as FMD may simultaneously act as a blow to the basic tissues of social life that damages the bonds attaching people together, but this may also *create* community, since shared trauma can serve as a source of communality in the same way as common language and common cultural backgrounds. Extraordinary events thus became anchored in and interconnected to everyday places in people's lives and FMD became a disease of the community.

Indeed the telling of the story, its context, highlighted common use of significant language that has specific meaning to the 2001 FMD epidemic. For example, there was much talk about being designated clean or dirty and of feeling dirty. Csordas (1994) argues that the body operates as an experiential base of biology and culture, thus in trauma victims, the physical reality of dirt cannot be separated from the emotional impact of the trauma. For many, the sense of being dirty stayed with them long after the muck was washed from their clothes. A slaughterman recalls how *you felt like lepers, you were shut off, you felt as though people didn't want to associate with you, not because of you personally but because they knew of the consequences... we didn't go out we didn't go anywhere.*

Here a woman describes going out from her culled farm: *I thought 'I hope I don't see anybody that I know of', because you felt dirty somehow, we would have been happier staying at home.* Again, there is resonance with the work of Hood & Seedsman (2004), who identified a powerful sense of stigma amongst farmers whose livestock tested positive to OJD, where farmers spoke of feeling like lepers or criminals and led them to withdraw from social contact.

So there was a collective FMD language. We have already introduced the war analogy, which was both explicit: *it's like we're at war but we can't see the enemy* and implicit, with common talk of pulling together; the killing fields, being culled out and rallying round. The FMD virus itself, so deadly and yet so invisible and intangible was also personified: *Foot and mouth was just like a demon.*

FMD narratives need to be understood in the context of shared experiences that are woven into everyday social relations and lifescapes. As Bearman & Stovel (2000:74) indicate, individuals tend to be embedded in relatively dense clusters of social relations in which the values that they hold and their sense of self are shared by and shaped by others with whom they interact. Thus shared FMD language may be used as a code or shorthand, that when understood opens up substantive or contextual institutional and cultural conditions that have shaped the experience and lived reality of the storyteller. FMD narratives may be an attempt to find meaning, to organize and make sense of extraordinary and for many, traumatic events. Shared language such as that described here, perhaps only makes sense to those who have also shared the experiences, experiences that such language attempts to represent. Indeed the 'insider/outsider' positioning was very hard to bridge. As a DEFRA field officer comments, *nobody can understand what we have been through unless they have been through it with them.*

Recovery

Post-FMD, tradition and continuity did return to the villages of Cumbria. In 2002 people were enjoying the chance to congregate again after a year in which the traditional social, sporting and religious calendar had been more or less cancelled with knock-on effects for local charities, businesses and organizations that often benefited from these events. The return to sports was therapeutic for some:

> *Usual busy week working, but managed an afternoon day sailing on Ullswater with my son and we came second in a race, first time on the lake since last year. Last year FMD almost ruined sailing season, due to ban on Lake, no excuse this year. Really enjoyed sailing again.*

A young respondent told us *I play hockey and towards the end of the hockey season* (in 2001) *that all got cancelled, because you know the pitches got closed.* In 2002 he wrote: *Now that the evenings are lighter it's meant that on nights off duty I've been able to get out more. It's made a very welcome change to be able to bike/walk on the fells again this year after all the restrictions of 2001.*

However, post-FMD confidence remains fragile, with a sense of collective dread when there are new FMD scares. Even happy occasions carry overtones of shared history, for example, a vicar reflects on a christening and how *every new beginning and every special occasion seems to bring with it FMD baggage, so small is our community and so heavily affected were we, that there is no escape. But I expect that it is all part of the healing process.*

Summary

Kai Erikson, speaking at the 'Voices of Experience' Conference in 2003[68], noted that collective trauma is the degree to which a line is drawn around the affected community that makes the people in it feel different from the people outside the line, and the people outside the line feel different from the people inside. He noted how that sometimes took place because the people within the circle come to think that they experience life in at least a somewhat different way than the people outside it; people within the line have shared an experience which sets them apart. Furthermore, the 'insider/outsider' positioning may have a lasting legacy:

> Thirty years on, the memories, pain and anger are still there. Incidents, small and large, can trigger off a memory. In Aberfan, memories of the disaster were brought back by news of the Dunblane killings. The parallel of the loss of the children inside their school was poignant and painful. Aberfan understood the pain of those who had lost their children that day in a way that few others could. [...] People who have not experienced such tragedies can not hope to fully understand the pain and trauma that disasters cause.
>
> (McLean & Johnes, 2000: 125–126)

In telling their story, people recounted the FMD disaster as it unfolded and was experienced within a particular *place*. Recent work by Wilson (2003) has examined the importance of exploring non-physical dimensions of 'environments', in particular those that do not exist solely 'on the ground' but are embedded within the belief and value systems of different cultural groups, placing emphasis on the social and spiritual aspects of place. Our understanding of 'place' is drawn from Sizoo (1997), who argues that place does not simply equate with community.

For many people in Cumbria, it was place in this broader context of taken-for-granted spaces, relationships and knowledges, which became disrupted and displaced by the 2001 FMD epidemic. As we have already discussed, the farm becoming a slaughter site on a massive scale, leaving cow byres *spookily silent* and the daily trip to work becoming a journey through *killing fields* with heaps of rotting livestock carcasses instead of lively, young livestock.

Boundaries between persons and things are osmotic and creative of one another, people, places and spaces are intimately linked, so that even when change occurs rapidly, people are always in some relationship to their environment they are never nowhere (Bender, 2001). Chapter 4 discussed in more detail how everyday 'materials' take on different meanings and how such materials can become displaced so that they become 'triggers' – igniting memories of the disaster. Harada (2000:206) writing about life in a public shelter, during the aftermath of a devastating earthquake in Japan in 1995, eloquently sums up the process of reconfiguring the social and the material:

> *What happens if the material 'order' or the 'social' meanings, which we take for granted, suddenly collapse and we try to continue our 'social' lives under extraordinary and materially different conditions. What we find is that spaces and materials need to be recreated and named, and that this process allows for the (re)creation of both collectivity and individuality.*

In the preface to this book, Kai Erikson describes how, just as the loss of a significant person in our lives can create a painful sense of also losing part of ourselves, so too we may experience emotions akin to bereavement when we lose something deeply connected to our sense of identity. 'This can be the case when we lose that in which we have invested much of ourselves…human beings have a sense of themselves which gets broadened by the things they own, the things they work with, the things they live with. And if those things are taken away, their sense of self is greatly reduced. It's as if a part of them is taken away.'

Another lasting change was how 2001 became *the* 'FMD' year, not just in terms of the disruption to daily life so graphically recounted or the time *lost* to FMD, but in particular as a dating device. This is discussed in wider disaster literatures, both in terms of a 'break', a fault-line between 'before' and 'after'. As Erikson (2003) notes 'They'll (those outside of the disaster) never understand how the disaster becomes a dating device itself. They'll never understand what you mean when you say, "well that was before." And also in terms of a qualitative difference between how those who have experienced a disaster and those who have not, date the event. The event is ongoing and it will not be over until the minds of those who experienced it reach a certain peace.'

Where does this take us in terms of reconsidering trauma? We would argue that in view of what people in disasters go through, traumatic stress is best represented as the 'normal' reaction of those people exposed to an abnormal disaster event.[69] In 2001 trauma was widespread, both acute and chronic, and characterized by feelings of shock, depression, including thoughts of suicide; loss of concentration and interest; recurrent thoughts and flashbacks, broken sleep, avoidance of and obsessive attention to details, problems requiring medical intervention, a sense of being misunderstood, and anxiety about the effects on children (e.g. witnessing culls). While such 'symptoms' accord with the clinical definitions of PTSD, we maintain that the experience of trauma in this context (while in many ways exacerbated by the long duration of the epidemic), should not be pathologized or seen primarily as a disorder. As López-Iblor (2005) indicates, there is a danger that those affected risk becoming victims, trapped by the situation, petrified in that position.

> A community nurse recalls *my worst memories.... taking my daughter and her friend home (from a show in Carlisle). It was the same evening that her father's pedigree sheep were being taken to the voluntary cull. By mistake I took them through a closed road – the sign having fallen down to the side of the grass – in the dark we went past a burning pyre only yards from the hedge separating the road and the field. We could see the charred, rigid bodies of the cows and the sparks from the fire and the smell permeating the air and the silence of the two young girls.*

Situations which may promote severe reactions are classically those where the actor or group cannot flee or fight the horror or terror, they cannot follow the instinct either to escape or confront the threat or danger. Being caught in such situations is believed to place the individual involved at risk of developing stress disorders. An example of such a situation involves a farmer who was prevented by movement restrictions from feeding his condemned sheep for three days and having to watch them starve in a field near the house. Many of these sheep were heavy in lamb at the time. Although the circumstances were particularly distressing in this case, this was not uncommon among those affected and similar situations have been related to us. Horrific as it was, these events took place within a collective context of distress, although of course it is very difficult to know to what extent this distress could be alleviated by sharing it with others. Individuals suffering distress, even nightmares, did not necessarily become 'ill' and many found their own psychological or social resources which offered them some protection against this.[70]

6
Working on the Frontline

Introduction

What do we mean by 'frontline workers'? In disaster situations, many people, some unexpectedly, end up working on the 'frontline' of the disaster. This is because the frontline often emerges in unexpected places. Disaster planning tries to predict these places and people, but disaster studies show that events cut across such plans and make heavy demands of people who get caught up in the chaos or its aftermath. In the FMD crisis frontline workers included field officers, slaughtermen, disposal site workers, vets, haulage staff. However, we could also argue that anyone caught up in the FMD epidemic was on the front line. These workers were often local people whose livelihood had been severely curtailed by the disease control strategies, who already had practical knowledge of handling livestock or the skills needed to operate the vehicles and machinery used in mass disposal; others were conscripted or seconded from a huge range of different occupations and parts of the world with very little prior training or preparation. They often had to operate in dangerous and highly stressful environments, to the extent that a number of respondents have likened this to 'war-work'.

Psychological problems and distress are frequently reported in emergency and disaster workers, and whilst stressful events are inherent in the nature of disaster response work, even situations that may be routine have been found to evoke stress among professional disaster workers (Secor-Turner & O'Boyle, 2006; Ursano et al., 1999). In one of the most frequently cited examples of disaster worker trauma, Duckworth (1991) describes several categories of psychological problems amongst police officers related to the 1985 Bradford City fire disaster in the UK, including performance guilt, anxiety generated from reliving the experience, generalized irritability, focused resentment, and motivational changes. More recently, however, Long et al. (2007) investigated American Red Cross disaster workers' symptoms of distress and posttraumatic stress resulting from exposure to disaster stimuli during their response to the September 11, 2001 terrorist attacks and found that 'the workers directly exposed to disaster stimuli reported no more distress than those who were not directly exposed.' What factors underlie such apparent contradictions in the literature? In this chapter, we highlight the issue of occupational stress and relate this to disaster situations and our study of FMD workers. We examine the response of disaster workers to trauma, focusing on what we term we prefer to call 'post-traumatic experience', and then we consider what helps and what doesn't. We develop an argument for the creation of 'therapeutic spaces' and peer-support networks for disaster workers.

Occupational stress and disaster workers

Perrewé et al. (2002) report that the literature on occupational stress and burnout has grown exponentially over the last 20 years. This interest is attributed largely to the recognition that stress can have detrimental effects on individuals' mental and physical health as well as negative effects on factors such as work performance and staff turnover. Put simply, stress is a fundamental element of the workplace that appears to affect adversely the well-being of individual employees (Dobreva-Martinova et al., 2002).

In the occupational health literature, research on occupational stress has frequently focused on role ambiguity and role conflict. 'Role ambiguity' has been described as stress resulting from uncertainty, often resulting from lack of clear job description, goals or specified responsibilities. The stress indicators associated with role ambiguity include depressed mood, lowered self-esteem, life dissatisfaction and low motivation.[71] 'Role conflict' refers to stress from conflicting demands, and occurs when an individual is torn by conflicting job demands or by doing things he or she does not really want to do, perceives as being not part of the job or suffers psychologically uncomfortable demands.[72] Hearns & Deeny (2007) note that relief aid workers who believed they were not being sufficiently supported in their original role could not cope with any new responsibilities, particularly where the resources did not match the workload required to achieve the outputs.

Recent work has highlighted the relationship between role changes, role ambiguity and post-industrial society, where labour markets tend to be associated with flexible production systems typified by adaptable, redeployable labour and high rates of turnover (Storper & Scott, 2002). For example, Hage & Powers (1992) discuss the links between increasing societal complexity in post-industrial society and relate changes in the structure of roles, identifying, amongst other things, the tendency for work teams to be viewed as temporary rather than permanent.

Issues of role ambiguity and role conflict linked to 'working on FMD' were apparent in 2001. In particular, those who were assigned the role of 'field officer' reported receiving limited role training and, initially at least, limited logistical support.

As one DEFRA field officer commented:

> *I was supposed to be at the farm gate handing out licences, but I was soon helping the vet hold down animals for slaughter, it was like that for the rest of the time...[people] didn't give any concern regarding their own health and safety. I think the events just took over and you just did what you could to help at the time. The role didn't really match the briefing at all.*

Another DEFRA field officer notes:

> *We had a choice to go dirty or stay clean. 'Dirty' meant going out to farms and help with slaughter, 'clean' didn't. I decided to go dirty, basically because I've worked on a farm, worked with stock, seen dead animals before so I knew it wasn't going to throw me. We all had to go, us 'dirty' people, to a training day up in Carlisle, where we'd be get told what our duties were...it was a complete and utter waste of time...it wasn't like that at all, I was off work with depression afterwards, I still haven't really fitted back into work.*

DEFRA field officers had a pivotal role in the management of FMD. They were the on-farm DEFRA contact, coordinating movements on and off farm, supervising the cull process and frequently becoming directly involved in penning and holding livestock as they were slaughtered. In many respects, for those already employed by DEFRA this was an obvious role for them to play as they would normally have close and regular contact with farmers and land owners over issues such as the management of DEFRA agri-environment schemes (for example, Environmental Stewardship and Environmentally Sensitive Area agreements). But as FMD continued and numbers of culled farms increased, many additional field officers were needed to cope with the crisis. They were recruited from different walks of life and included those whose usual employment was impossible during the epidemic, for example fencing and landscape contractors. A farmer who lost livestock (through dangerous contact regulations) during FMD comments on the role of the FMD field officer:

> *The field officer's first job was to go through all the paperwork with the farmers. They seemed to have to fill in a lot of forms before slaughter started. They were meant to co-ordinate operations at an individual farm and make sure that things were done in the correct way. I think this meant seeing vehicles on and off, presumably counting heads of stock, making sure vehicles and people got disinfected etc. The job varied a lot depending on the person and the situation. Some seemed to stand at the gate with a clipboard, others got stuck in and helped with slaughter arrangements and also looked after animal welfare. As with so much, we never got definitive information about what field officers were and what was their role. We never saw the same field officer twice although eventually there was one who was meant to be in charge whose phone number we were given, the number was wrong, we eventually got the right number but I don't think we ever managed to speak to him. I think for every farmer who had a field officer, or succession of field officers, you would get a different story* [about their role].

The literature on role conflict suggests an association with reduced job satisfaction, higher anxiety levels, psychological strain and physical health problems such as heart disease. It is also strongly linked with burnout (Perrewé *et al.*, 2002; Chang & Hancock, 2003), which is typically defined as a syndrome of emotional exhaustion and depersonalization as well as an inability to cope with stressful situations. The result may be a sense of hopelessness, fatigue, and range of psychological and physical health problems and deterioration of family and social relationships.[73] Here a DEFRA field officer describes withdrawing from social contact at home:

> *I was coming home* [very late at night] *and I didn't want to talk to anybody. I was just ignoring* [partner] *I wouldn't talk to him. I wouldn't phone my parents, I wouldn't phone any of my friends. I just wanted nothing to do with anybody.*

A study of the Piper Alpha disaster highlighted that the factors which appeared the most powerful causes of stress, were principally organizational and managerial.[74] For example, paperwork, lack of personal recognition, obstacles at work, lack of opportunity to display initiative (Alexander & Wells, 1991). Hearns & Deeny (2007:34) also highlight how a 'lack of resources in the field was a constant problem' and discuss the stress associated with a perceived 'lack of understanding from management in headquarters of the reality of the situation.' They quote one respondent who felt that there was an 'unreal expectation by HQ; we have too many extra role responsibilities.' Correspondingly, during the FMD crisis, workers at 'Portakabin City',[75] reported frustration at not being able to carry out their roles because of perceived excessive bureaucracy:

> *It was like trying to dress an octopus in a set of bagpipes. They* [DEFRA headquarters] *wanted to work in traditional ways* [regulations], *we were working like farmers do, making deals and negotiating ...* [DEFRA headquarters was] *an abyss into which things go and never come out.*

> A vet told us that *initially it was a bit frustrating because they were clamouring for vets and then I got there at nine o'clock one morning and sort of sat around 'til one o'clock. it was quite chaotic in some ways, I remember feeling very frustrated at times about the system. Quite a lot of the time you weren't told specifically enough what, they wanted you to do. I think man management wasn't good at all, I mean they're very quick, quick to criticise. There was a meeting every the morning and again, without sort of naming names at the top of DEFRA there are certain people who are very quick to criticise publicly, but never gave any praise publicly.*

Another DEFRA field officer notes:

> *The main event of the week regarding FMD was the DEFRA meeting on Friday morning. Each department spoke on its role in FMD and more or less congratulated itself on how 'wonderful' it performed. It was almost amusing hearing some departments...*[some] *presentations I thought were a bit misleading.*

There is evidence that the risk of role stress and trauma in disaster workers may be reduced by pre-event training and preparation.[76] For example, Brondolo *et al.* (2007:9) argue that steps can be taken on 'both an organizational and an individual level to improve the degree to which the risks associated with disaster events are appropriately anticipated and managed...we can and should provide safety procedures' [for frontline workers]. Other work has highlighted how the notion of 'victim' is rarely used in relation to rescue, medical and support workers (Alexander & Wells, 1991; Shepherd & Hodgkinson, 1990). Disaster workers are often considered to be resourceful and impervious to traumatic experience.[77] Yet Deaville & Jones (2001:7), commenting on the sustained mass killing experienced by many frontline workers during FMD, question if training or experience could provide adequate preparation for such work. Here some frontline workers emphasize the scale and nature of the mass culls and the incongruity of the work. Below a worker involved in the transportation of animal carcasses reflects on the difference between the work at the beginning of the epidemic, when carcasses had been lying for days following being culled, then later, when disposal was much more rapid.

It got to the stage, where we'd give the farmers a hand, throw them in the buckets, there was bits and pieces everywhere, you just had to get on with it. If you'd tried to get everything on the loading shovel you'd never of got anything done. People going across two fields to pick sheep up, you know, one here and one there and leg here and bit here and bit there...you just got on with it. Then later, *at the end, you were sort of going, and we were maybe waiting for the slaughter team to come, and as soon as they shot them we were loading them which for the smell side of it, it was a lot better, you know. But it still wasn't very nice when one minute you see them all running about and the next minute you're just putting them in a wagon like.*

Similarly, a disposal worker noted:

When we first started, they'd been shot and they'd been left for ten, nearly fourteen days, some of them, and I wouldn't have anything to eat for the first three days, you just couldn't stomach it, then after that, queer thing to say, but you got used to the smell. I used to come home and I could hardly smell it on us and as soon as I walked through the door I just stripped off. It was absolutely horrendous like, to see it, it used to just cling to you. But it's just... If I hadn't done it, and if nobody else had done it, the job would never have got done. He talks about the waste: *I went to a lot of farms and I was loading them and they were shooting them and they didn't even have foot and mouth. You know, I mean I'm talking farms were they had seven, eight hundred cattle, three or four thousand sheep, well to me that was just absolute sheer waste. I think the farmers would have accepted it more if all the ones that had been clean had gone to the butchers and the public had got something out of it. Like the way things are now with foot and mouth, you know, the way it's been* in Britain, we should have been eating beef for end of our days, instead of just shooting them and just wasting them. I've been to farms where I've loaded over 4,000 animals and there was nothing wrong with them. I've been to farmyards where there was that many cattle in I couldn't even turn the loading shovel. They were that deep, and the farmer was standing there saying there's now't wrong with them.

A field officer also reflects on the waste:

Well there was just wagons rolling in, you were just checking the wagons, one after the other. It was just constant, one wagon after the other, same old smell. You just didn't feel like you were doing a decent job. It was just drudgery, eight hours a day or more. I just couldn't see the point to it. It just seemed such a senseless waste of all these animals. That was the biggest impact on me. At least on the farms it was terrible, but I felt as if I was doing a decent job and I was helping.

A number of workers found that they just couldn't cope with the job.

The public didn't have a clue what was going on. I mean if they had to go and do what we'd done...One lad went on a loading shovel, he got half way down the lonnin,[76] he jumped off and he says 'I'm away', he says 'I can't hack this job'.

Another worker involved in haulage discusses how the scale of the culling meant that there were pressures to kill animals quickly and then move onto the next farm. Inevitably: *one or two slaughter teams didn't do the job right*, often resulting in animals suffering. But even on the grimmest of jobs workers struggled to uphold their personal standards and cope with the horror. They did not want to compromise their desire to do a job properly.

They'd shot them but they weren't dead, and I just... I did have an argument one day with one bloke, and I said, sorry but I go shooting every week, and I says, that cow there is not dead, I said if it's dead I'll load it, and I just refused to load it. This fella in the wagon, he's shouting at the army bloke and the vet and they're arguing away, I just says I'll load when they're dead. He said they're just animals, I said I don't care what they are, they're the farmers livelihood, he's standing ower [over] there watching, he stands there and watches them get shot properly, and he's satisfied, it's got to be a little bit of sympathy for him. At the end, 20 minutes, half an hour later, they

> had to come and shoot three of them again. There was one in the wagon. They had to climb in the wagon again and shoot one, and there was two on the floor, and they weren't dead like, they were breathing.

For most frontline workers, daily conditions of work changed substantially. People spoke of long hours, exhaustion, sporadic meal breaks, and keeping going on adrenalin:

> Well we didn't have a social life because you were just going to work and coming back and that was it. You know, you were coming home and if it were that late, you were just coming home, stripping off, walking through, having a shower. Half the time you didn't have anything to eat because you just went to bed, you didn't really want now't to eat.

A slaughterman recalls:

> Five or six weeks later when we realised how long we'd been working because when you're working all the time, job to job to job you don't realise the obvious if you know what I mean. You were driving up and down motorway and you used to count pyres, and when there started to get above 20 pyres burning at once the penny dropped and then the depression sets in cos then you're, you're just slaughtering [the] future.

Disturbed sleep was widely reported, another member of a slaughter team recalled:

> I think the biggest day I done was 575 dairy cows – it took 17 hours… then you lay down you couldn't sleep… You only used to sleep two hours then you'd wake up… you were just on autopilot all the time.

People working on the frontline who lived in non-infected areas would often not go home for weeks but stayed with friends or in lodgings. A slaughterman (with a young family) who lived in a farming area said:

> I went on the 13th to this farm, she took us and dropped us off and I was never back home for eight coming nine week, I went and stayed (in town) with a lad I worked with so… That was nine weeks I stayed away.

A field officer simply noted that: *It possessed your life, it really possessed your life.* Wagon drivers faced new imperatives when collecting milk or delivering feed and were under constant scrutiny regarding bio-security.

Here a milk haulage manager recalled his sorrow and stress as customer after distraught customer telephoned to let him know that they had lost their herd and how after listening to their stories he struggled to re-arrange his tanker collection routes to avoid the infected farms and the roads around them.

> *And [the farmer] rang me up first of all and I'll be honest with you, I shed tears with him. And then of course I had to ring head office to tell them and there was one of the girls I spoke to, she was always quite a friendly girl over the phone and very understanding. And I think both of us was in tears before I finished with her, you know. You got a phone call one after the other. And each time there's a phone call came well, you had to start and re-hash the routes which wasn't easy. Because I mean, you'd have an area where you'd have you know, your two kilometre area what have you, the farms was classed as 'A' farms and you had to pick them up last on the route. Well quite often, I mean, officially I should finish sort of 5, half past at night, quite often I was there at half past 7, half past 8 and 9 o'clock. You'd just be ready to go home, they'd be three or four phone calls you had to start and rehash the loads again.*

The scale and cost of the eradication was particularly divisive. For example, a vet notes that one of his clients thinks that he made a lot of money during FMD: *The farmer is convinced he was given FMD deliberately and on arrival I was given his weekly tirade regarding DEFRA, Tony Blair, how I must have made thousands of pounds out of it etc.* Yet some vets, like this one above, were seconded from their practices who paid their normal salary, the balance of the DEFRA fees going to sustain the practice as a whole. Variations in salaries paid during FMD led to market distortions which created resentments. A slaughterman says of a former fellow worker: *He phoned up yesterday, and they're still on standby from DEFRA. Never done a day's work since September and he gets £220 a week.*

The bio-security measures, of microscopic disinfecting and 'washing out' of infected premises for weeks following a cull, were frequently described as punitive: *You started thinking of all sorts of strange things in your mind washing out. I said prisoners wouldn't have done it.* Severe movement restrictions affected a wide range of rural workers. Community nurses report having to take medical histories and make assessments:

> *Over a gate...there was a siege mentality in all aspects of life...my job changed significantly during the foot and mouth epidemic...the guidance from the practice changed all the time, I found myself discussing bowel*

problems with a farmer and having to shout over the farm gate. I got very angry.

Again, there are striking parallels with the account of the process of the OJD outbreak in rural Australia. Hood & Seedsman (2004) state how the government staff witnessing or conducting the slaughter were traumatized and were subject to open hostility, exhaustion, and burn-out. Social psychologist Joseph McGrath's (1976: 1352) often-cited definition of stress highlights the significance of perceptual factors in the individual as a significant factor, stating how the 'potential for stress exists when an environmental situation is perceived as presenting a demand which threatens to exceed the person's capabilities and resources for meeting it'. The following contributions from a field officer seconded to the crisis from the Environment Agency underline this reaction:

I kind of switched off... I just thought this is absolute chaos, this is madness.

He felt morally compromised:

I resented myself and I resented, the Government and the fact that the only way I could comfortably resolve the situation was to, leave my job.

And suffered extreme anger:

I vividly remember, the last farm I was on...I was penning up cattle and, I would have been quite happy to have seen (politicians) getting penned up and getting popped in the head...which I don't think was a very healthy state of mind to be in, but...that's how I felt at that time.

He speaks of flashbacks and suicidal thoughts and recurring traumatic images:

I'll never be able to look at a cow or a sheep again without seeing blood pouring out of the hole in its head, maybe I will in time...I walked, walking along the pier one night. I did actually think about jumping in... I felt so bad about myself.

Another field officer stoically noted:

You were doing something, it was just a job, you just had to get on with it. If you started bursting into tears, breaking down in

> the farm you'd have been no use to anyone, you just had to keep going.

While many front line workers were local people whose livelihood had been severely curtailed by the FMD control strategies and who had practical knowledge of handling livestock, others were seconded from government agencies or unrelated branches of the Civil Service. Another field officer told us:

> *They were literally taking anybody on but some people had no knowledge of livestock whatsoever...because, as far as they were concerned, we would be standing, in full waterproofs with a clipboard, at a gate directing traffic basically, licencing folk in and out, making sure they were disinfected properly, basically getting in the bloody road!*

But being a field officer involved much more than directing traffic and paperwork.

> *I mean, in between all this* [culling of cattle] *I was keeping an eye on the sheep and I was lambing some of the sheep!* [the sheep were unlikely to be culled until the following day] *I know it sounds completely silly, but I just couldn't walk past and, and leave them.*

Again, the analogy of war is used to describe working on FMD as a former soldier recounts what it was like to work on the slaughter teams.

> *I've counted over the last six months how many cattle and animals I killed ... I can only relate to it as like being in the forces, like going to war.*

Here a slaughterman comments on the role of the field officer and the *difficult situation* they were in, both in terms of managing the day to day operations on-farm and disclosing information about the management of the epidemic:

> *I says, 'Oh you've likely got field officer bible read then eh?' Aye they couldn't say 'owt out of turn or anything that would incriminate them.*

Some felt that local knowledge and direct farming experience would inure the workers to the hardships of what they had to do, others were not so sure, particularly in relation to some of the (younger) farmers sons involved in the culls:

> Richard is actually a farmer's son, and he found killing, that the killing, the disposing of the animals, very hard., I think the ones who managed to keep on going, were the men who hadn't looked after animals, mebbe country born and bred but not having roots directly with farmers. They managed it but the [young] lads like Richard.... they couldn't hack it, and there's an awful lot of them. I think there's a lot more going to happen, mental problems for a long time. Oh yes, I mean Richard doesn't like to talk about it.

Long et al. (2007:306) identified age as an important factor in their study of 9/11 Red Cross disaster workers, and commented that the workers, with a mean age of 56.8, were on average 20 years older than other '9/11' workers. They note that younger ages in disaster workers have been found to be associated with higher rates of trauma and PTSD symptoms (see also Fullerton et al., 2004; Gaher et al., 2006).

From their study of FMD in Wales, Deaville & Jones (2004) state that many of the frontline workers found their inability to help distressed farmers debilitating and emotionally taxing. A field officer relates how they were frequently asked: 'why is MAFF doing this, and you haven't got a answer for everything, but they expect you to have answer because you're working with the organization you're supposed to know everything.' Kumagai et al. (2006) argue that disaster agencies and workers are often blamed when assistance fails to meet public expectations, and a number of field officers report being threatened or verbally abused by farmers:

> I cried on most sites and vomited on one. I was hit by a car, knocked out and abused by some farmers.[79]

But whilst there were sometimes very stressful situations on farms the fact that many field officers were local people meant that there was often a good understanding on-farm:

> We also built good relationships up with some people, there was a woman [farmer] up near [village], she made 150 cups of tea and chicken and chips for 37 people, it was amazing, she was crying all the time she was doing it.

A slaughterman also recalled:

> [There was] *very little anger towards us, quite a bit of sympathy towards us to be quite honest. You know we must have looked pretty down in the mouth you could tell you know, way people reacted to you. I must look pretty depressed, a lot of people obviously felt a bit of sympathy to us. 'Cos obviously they were having that one disaster on that one farm but we'd just done 50 farms beforehand and we were going to face another 50.*

Additional evidence regarding the health problems experienced by DEFRA frontline staff was provided by a questionnaire survey completed by (the then) DEFRA Occupational Health Department in 2002 after the crisis was over.[80] The questionnaire was sent out to all staff that had worked on FMD, with 399 completed questionnaires returned (response rate unknown). Questions included location, length of time staff worked on FMD-related activities, and type of activity (diagnosis, slaughter, burning and burial). Staff were also asked if they had experienced health problems (the questionnaire included questions related to respiratory problems, vomiting and diarrhoea, blisters and stress or mental health problems, as well as an 'Other' option). In total, 28% of respondents reported work related illness. The most frequently reported health issue was stress or mental health problems (14.5% of workers), followed by respiratory symptoms (8.7%) and vomiting and diarrhoea (7.3%). In addition, more than 14.5% of workers reported an accident whilst working on FMD, but only 9.5% actually reported accidents at work during this time. For the 'Other' category, self reported conditions ranged from 'sore throats' and 'fatigue' through to 'post traumatic stress'.

Of course 'in-house' surveys of this kind should be treated with caution, not least because staff were asked to give personal information on the questionnaire, which would have made them easily identifiable within the organization. Nevertheless, the fact that 14.5% of DEFRA staff felt able to disclose to their own occupational health department that they had experienced stress or mental health problems is telling, and gives some indication of the highly stressful conditions workers endured during FMD. This raises the issue of ongoing (psychological) trauma as a result of these experiences.

What helps and what doesn't

There is much literature concerning the effects of 'stressful or traumatic work' on mental health and emotional well-being and how best to manage it.[81] A key issue which gets discussed is the significance of colleague or peer support. Dryden & Aveline (1988), for example, identify that institutions and groups with strong esprit de corps foster feelings of loyalty and a sense of belonging. Such support can have a positive, healing function, particularly when workers have experienced traumatic events. Hearns & Deeny (2007) note that a perceived lack of support led relief aid workers to express low evaluation of self and feeling bad about how this affected their performance.

Conversely however, Paton *et al.* (2000:177) note that 'while membership of a cohesive group generally enhances resilience, it can, under certain circumstances, have the opposite effect. For example, cultural characteristics which advocate emotional suppression can increase stress vulnerability.'

Group membership may also act as the focus for negative coping mechanisms such as alcohol misuse, smoking and drug use. Davey *et al.* (2000) highlight the link between occupational stress in the police service and alcohol and drug use, and associate drinking subcultures with good teamwork, resulting in peer pressure to drink (Fillmore, 1990, cited in Davey *et al.*, 2000). They note that 'the working culture of the police service appears to prevent officers from adopting more appropriate coping mechanisms such as social support, counselling, stress management strategies and exercise.' Violanti (1993) also identifies the link between alcohol use and occupational stress in the police service (see also Violanti & Paton, 1999; Violanti & Aron, 1993). Long *et al.* (2007:306–307) suggest that the apparent lack of distress amongst 9/11 Red Cross workers may be because, amongst other things, those workers were 'well educated' and 'more likely to be married' both factors associated with 'improved socioeconomic status, engagement in generally healthier behaviors and built-in source of support.'[82]

The value of more formalized debriefing interventions is even less clear-cut.[83] Everly *et al.* (1999) carried out a meta-analysis of the effectiveness of psychological debriefing following trauma and reported results which supported the effectiveness of group psychological debriefings (though with the caveat that further work was needed in relation to the nature of the intervention programme). Such findings are supported by other studies covering a variety of disaster scenarios, including the Los Angeles riots[84] (Hammond & Brooks, 2001) and rescue workers following a major earthquake (Liao *et al.*, 2002). Bisson *et al.* (2003:145) reviewed the evidence base for psychological intervention following traumatic events and concluded that 'brief cognitive – behavioural early intervention may be beneficial and should be primarily aimed at individuals with acute symptoms, as opposed to everyone involved in a traumatic event'. Hammond & Brooks (2001) argue that police officers and fire-fighters receiving as little as 1.5 hour debriefing within 24 hours of an incident exhibited statistically significant less depression, anger and stress-related symptoms at three months than did non-debriefed subjects.

There is, however, contrary evidence suggesting that psychological debriefing is ineffective and can impede natural recovery processes (Raphael *et al.*, 1995; Mayou *et al.*, 2000; Schouten, *et al.*, 2004). According to Carlier *et al.* (1998:143) the findings of comparative studies regarding the clinical effectiveness of debriefing were disappointing. Indeed, 'eighteen months post-disaster, those who had undergone debriefing exhibited significantly more disaster-related hyperarousal symptoms.' Critics of debriefing argue that formal debriefing interventions may pathologize normal reactions to abnormal events and thus may undermine natural resilience processes (Bonanno, 2004).

Alexander & Wells (1991) reported (regarding the Piper Alpha disaster) that post-traumatic reactions such as intrusive images may be reported without there being raised levels of psychiatric illness or increased sick leave. They found that officers conducted their duties successfully (body retrieval and ID) and 'emerged relatively unscathed, and that some even seem to have gained from their experiences' (Alexander & Wells, 1991: 551). Wilson (1980) also describes how emergency services staff can become addicted to the experience of trauma and may struggle to function effectively without it. Hearns & Deeny (2007) highlight the relative importance of the 'team' to frontline workers. In a study of relief aid workers, they note that the team composition was mentioned many times in various ways, such as team spirit, cohesiveness, and the ability to identify with the others. Working in a team was likening to a 'family structure'.

The notion that frontline workers may gain positive benefits from disaster work may appear at face value difficult to understand, but the experience of 'team work' and 'helping' provided strong emotional and psychological support for many frontline workers during FMD. For example, a DEFRA field officer speaks of *doing a good job, helping in some way*. He liked the *adrenaline buzz of working flat out*, now he feels *flat and dissatisfied with my life*. Another field officer recalled: *a good team... I think there was actually* a real sort of...you know, at the end there was a real sort of comradeship about it.

Some frontline workers also felt aggrieved that their hard work and efforts had not been acknowledged:

> Had my job review done, not happy with it, hardly mentioned the FMD work last year. It is almost as if last year didn't happen. Similarly, I suppose it's human nature we want some acknowledgement that we did something during that time and we haven't had that, we haven't received that [...], its not just acknowledging what we've done, its acknowledging that there was a problem and a lot of people are still in sort of denial to it really I think.

Disaster 'social capital'

So what to make of all these contradictions? We want to argue that 'off-the-shelf' plans to address all needs are unlikely to be effective given that disaster situations are multiple and dynamic. Rather, what is required is recognition of the sometimes hidden, but extremely important networks that form and coalesce around specific disaster events, these might be called the 'social capital' of disaster work teams. Such networks are important not only in terms of managing the disaster, but also exist as ad hoc support structures for those involved in the situation. Despite the doubtless well-intentioned interventions of DEFRA and other agencies to 'debrief' their staff following FMD, what was offered was essentially a formalized, 'off-the-shelf' service, which did not reflect the experiences and needs of frontline workers, as a DEFRA field officer notes:

> What was DEFRA like in terms of the support they offered? Pretty nonexistent...nobody can understand what we have been through unless they have been through it with them. The counsellors would probably talk you through divorce, separation, bereavement whatever, but what we went through was way above any of that.

In contrast, peer support and informal counselling, happened spontaneously, facilitated through established networks of trust:

> I did counselling, counselling for 2500 people who came through my door! There were vets from all over the world, they'd come in and have a cup of tea. We didn't use the counselling service, we'd counsel each other. I'd sit

> in the rest room, there would be about 15 of us, and we'd just talk about the day's events, some of those farms, it got nasty. It could be violent.

North & Hong (2000) highlight how during disasters, people seek support from trusted members of their own communities rather than mental health professionals, and advocate the training of volunteers to support their peers. Importantly, they state that distress is commonplace following a disaster and that most people are not psychologically ill. Following terrorist situations in Israeli in 2002, a study of 'frontline' nurses (Riba & Reches, 2002) found that they wanted group support sessions with their peers soon after an incident. The debriefing did not need to be organized, and included simply sharing a cup of coffee, but many nurses pointed out the importance of 'some time-out' with their peers for mutual support. Jonsson & Segesten (2004) also argue that through dialogue with colleagues, it is possible to cope with the effects of traumatic experiences, highlighting in particular the importance of making the experiences understandable and meaningful. It is salient also to emphasis that peer support, rather than supervisor or managerial support, is important. A study of ambulance workers (Hedlin & Petersson, 1998, cited in Jonsson & Segesten, 2004) identified that whilst good support from peers was vital, less than 50% of workers thought that they could talk to their supervisor or manager. Alexander & Wells (1991) identify the importance of good relationships amongst police officers involved in body retrieval work (following the Piper Alpha disaster), and found that humour and talking with colleagues were the coping strategies police officers found most helpful in dealing with their work. Most recently, Alexander (2005:13) has highlighted revived interest in 'psychological first aid', which includes principles such as 'protection from further threat and distress', 'helping reunion with loved ones', 'sharing the experience' and links with 'sources of support'. Based on the experiences of frontline workers during FMD, we would support such perspectives. In disaster situations, agencies need to create 'therapeutic spaces' for workers to talk over experiences, opening opportunity for dialogue channelled through existing trusted peer-support networks.

But it is also important to note that for the *temporary* workforce of frontline workers co-opted or recruited to work for DEFRA at short notice (from all over the UK) and then often disbanded just as quickly, formal counselling services were not available to these staff (or, in all probability, the informal networks we mentioned above). Agencies have a responsibility to all frontline workers called upon in disaster situations, not just permanent members of staff.

In addition, frontline workers amassed huge expertise during FMD. We argue that this is still not being sufficiently recognized or recorded and thus cannot inform future contingency planning (as identified by the Cumbria Inquiry, 2002). For many frontline workers, this formal lack of recognition has compounded a prevailing sense of anger and frustration. Agencies that employed front-line workers should have tried to capture and record these acquired skills and experiences further. Ensuring ways of accessing such knowledge by calling upon individuals and groups should be part of contingency planning, this would not only formally recognize the contribution of frontline workers in managing the disaster, but also bring together strategic and operational knowledge about the experience. This would be a critical and trusted resource for the management of future outbreaks.

Summary

Many people who worked on the FMD epidemic were not emergency or disaster workers in the traditional sense. Nevertheless they had to deal with difficult and highly stressful situations, which they often endured on a daily basis, for weeks if not months. The evidence suggests that during disasters, frontline workers tend to rely on their informal sources of support or 'social capital' and seek support from trusted colleagues, who are able to 'share like experiences', rather than more formalized support from mental health professionals or counsellors. As a DEFRA field officer indicated, *you can only appreciate it through having experienced something similar, and I think for a long time I was looking for that kind of support.* In disaster and post-disaster, there is a need for staff to support each other and organizations to encourage and facilitate such seemingly 'low-key', yet vital, 'therapeutic spaces'.

7
Exploring the Lifescape

Introduction

The term 'lifescape' refers in this study to the complexity of the spatial, emotional and ethical dimensions of the relationship between landscape, livestock, farming and rural communities. As discussed in Chapter 3, the mass slaughter, often of healthy animals taken out under the contiguous cull, 'dangerous contact' rationales, or simply misdiagnosis, was all the more horrific because of its being out of place and out of time. In this chapter we consider the concept of lifescape and offer a framework to explore the lifescape changes enforced by FMD.

Origins of the lifescape

The concept of lifescape has evolved within social anthropology[85] as a way of framing the complex relationship between people, place and production system. Lifescapes are necessarily interactive; people and places are intimately interconnected (Howorth, 1999). As such, lifescape links with a substantial body of literature which explores the relationship between people and place. Authors have, for example, focused on the social construction of place, how place meanings develop over time and how people become attached to place.[86] Teo & Huang (1996) argue that places are not abstractions or concepts but are directly experienced phenomena of the lived world. As Doreen Massey (1994, 2002) argues, rather than thinking of places as areas with boundaries around them, they can be imagined as articulated moments in networks of social relations and understandings. She questions the meaning of 'local place', arguing that it has become synonymous with other terms like everyday, real and lived. She states that if place really is a meeting place, an arena for negotiation and discussion, then the lived reality of our lives is far from being localized, as the connections, sources and resources of 'daily life' spread much wider, and asks where would we draw the line around the grounded reality of the everyday?

Our use of the term 'place' in this book encompasses everyday understandings of home and location, but also as Teo & Huang (1996:310) indicate, place is an 'active setting which is inextricably linked to the lives and activities of its inhabitants.' Thus place is also about situated social dynamics and is multi-dimensional, holding different meanings for different social groups. More normatively and linking with ideas about therapeutic spaces, it also comes to mean safe spaces and settings for health and health practices. For those living in the areas worst affected by FMD in 2001, familiar surroundings changed and became hostile, bleak, beleaguered, and (from their perspective) unhealthy and were described as: *unreal, bizarre, crazy, and surreal.*

Very little has been written about the theoretical frame of lifescapes, though it is clear that the concept exists in other discourses in other forms. It has, for example, meaning in the transformations of social space which Bourdieu (1977) refers to as the habitus and in the phenomenology of perception discussed by Merleau-Ponty (1962). It can also be found in the *lifeview* literature of social archaeology (Bender, 2001; Robin, 2002) and the emotional landscapes literature of Gesler (1992, 1993, 1996), which emphasizes the importance of places for maintaining physical, emotional, mental and spiritual health. In geography, the lifeworld (Buttimer, 1976 and later Seamon, 1979) has been used as means of drawing together the phenomenological with the existential to bring new meaning to emerging concepts of humanistic geography (Daniels, 1994). Buttimer's influential 1976 paper 'Grasping the dynamism of lifeworld', argues that much of man's *(sic)* relation to nature is a function of 'human constructions of life, value, health and rationality...patterns of living shaped by the sense of space.' Indeed, time-space geography has consistently explored the contextual relationships between people, space and meanings (Robin, 2002). The first part of this chapter aims to explore the antecedents and meanings of lifescapes within some of the above-mentioned literatures.

Therapeutic landscapes and everyday places

In articulating therapeutic landscapes, Will Gesler (1993), highlights how place influences personal identity. However, much of the literature on therapeutic landscapes emphasizes the importance of symbols and symbolic landscapes in shaping health (Kearns and Barnett, 1999; Wilson, 2003). There is a need to consider the therapeutic value of everyday geographies, the culturally specific dimensions of the links between well-being and place, and the significance of place in people's everyday lives. Within the last five years and particularly within the geographies of health, such work is beginning to emerge. For example, a mixed method study of four localities in two cities in the North West of England explores the links between place, health related social action at the individual and collective level and lay knowledge. In so doing, the study questions how shared social meanings about a place, described as 'normative dimensions', may shape the way people respond to the everyday experiences of living in a particular place (Popay et al., 2003). This notion of everyday place is fundamental to lifescapes. The 2001 FMD disaster dramatically changed the lifescapes of study respondents certainly in the short term. In the longer term, it is almost impossible to calculate the ongoing resonance of FMD, not least because of the individual, nuanced nature of lifescape disruption.

FMD also highlighted the heterogeneous complexity of the rural landscape. Doreen Massey's (2002) multiplicity of trajectories approach to place and community is helpful in that it serves to move the 'action' away from site-specificity towards places that work, and are worked, on many different scales, emphasizing place as 'a doing' (Rose, 1999; Bender, 2001). It articulates what Tim Ingold (1992, 2000) refers to as the being-in-the-world attachment to place and landscape, highlighting that through familiar fields and woodlands, roads and paths, people create a sense of self and belonging. Our ways of *knowing* are thus strongly linked to our ways of *being*. Similarly, Lee (2007) argues that the relationship between humans and landscape needs to be understood as experiential, involving the subjectivities of people. It is through interactions/interactivity with the land people shape their surroundings according to intentional patterns of use in the future as well as the past. In a farming context, this complexity is not easy to articulate, it is at once ubiquitous and specific, open and hidden, and deeply embedded in the nexus of landscape, livestock and farm. As Gray (1998) notes, this consubstantial relation is not created and known through [farmers'] self-reflexive contemplation or theoretical discourse about their farms. Instead, it is the outcome of their everyday farm work.

Lifescape, lifeworld and habitus

The concept of lifescape has resonance with the sociological theory of lifeworld, which originated in German phenomenology as *lebenswelt*, signifying a relationship of intentionality between a conscious and imaginative human subject and the external world (Johnston et al.,

2000). It is the everyday life, anchored in a past and directed towards a future, a shared horizon, though each individual may construe it differently (Husserl, 1978). The lifeworld, according to Habermas (1987:124), is the fabric of everyday life represented by a culturally transmitted and linguistically organized stock of patterns of interpretation, valuation and expression. The lifeworld reaches right into the heart of social reproduction as communication and facilitates what is necessary for social life to exist. Sayer (2000) states that lifeworld is a product of the relation between embodied actors and cultures into which they are socialized, and draws on Bourdieu's habitus as a major influence[87].

Terry Lovell (2000:12) provides a clear overview of habitus, writing that an individual's habitus is not a matter of conscious learning or ideological imposition, but is acquired through lived practice. It is manifested through the 'practical sense', the ability to function effectively within a given social field, an ability which cannot necessarily be articulated as conscious knowledge: 'knowing how' rather than 'knowing that'. Habitus is deeply ingrained habits of behaviour, feeling, thought. It is ways of doing and being.

The dynamism of habitus is later expanded by Bourdieu when he writes that the habitus in a society is not uniform and differs, both between societies and within society, by educational background, fashion and social pressure. Frankel, a social archaeologist, writes that a key aspect of the concept of habitus involves the process of individual enculturation and its implications for social reproduction; habitus thus refers to all aspects of society, as lived through everyday experience (2000:168). In his 1977 Outline of a Theory of Practice, Bourdieu defines habitus as a generative system of 'durable, transposable dispositions' that emerges out of a relation to wider objective structures of the social world. 'It is the dialectical relationship between the body and a space structured according to the mythico-ritual oppositions that one finds the form...which leads to the embodying of the structures of the world, that is the appropriating by the world of a body thus enabled to appropriate the world'.

Sociologist Mick Smith (2002:3) writes that the transformations of social space central to Bourdieu's habitus are fostered by particular social practices and the adoption of particular (discriminatory) dispositions towards one's surrounding environment, including the relationship between man (sic) and the natural world (Bourdieu, 1977:91). The habitus is a form of practical sense operating without the necessary mediation of conscious thought, the habitus becomes 'second nature'. It is a (necessary) coping-mechanism that enables us to respond immediately and 'appropriately' to the circumstances of everyday life.

Bourdieu's work also provides an important critical perspective on phenomenology, an engaging account of which is provided by Throop and Murphy (2002), who argue that Bordieu's position is centred on a critique of subjectivism – objectivism: the subjectivist orientation of phenomenology concentrates too heavily on the immediate experience of the individual and their own interpretations of the social world. On the other hand, objectivism fails to take account of individual actors' actions, and instead locates them within the social framework, tied to objective relations of social structure. Bourdieu's trick, they argue, is in merging these two traditions and using habitus to explain individual experience while still retaining a view of the world as a set of relatively obdurate objective structures. Similarly, Wacquant (1992: 9) comments that Bourdieu criticises the phenomenological focus on descriptions of lived experience in an attempt to make explicit the 'primary experience of the social world', phenomenology is limited to apprehending 'the world as self-evident', 'taken-for-granted'. In other words, phenomenology is flawed because it does not explore how lived experience is produced through the dialectic of internalisation of previously externalized structures.

There is a counter argument, however, that Bourdieu simply wants to have his cake and eat it. Throop and Murphy (2002) argue that Bourdieu often trivializes or simply discounts the meanings attributed to events, experiences and behaviours that are given by the people themselves. They state that Bourdieu's emphasis on the non-conscious grounding of social action leaves little room for giving validity to the world as it is experienced by actors. Indeed, Frankel (2000) locates habitus within the more 'mundane aspects of life' and refers to habitus as 'everyday practical behaviour', thus placing habitus within a broadly phenomenological frame. Mouzelis (2000) also argues that in so far as Bourdieu stresses the practical character of habitus, the fact that it can be very flexibly applied to the ever-changing context of the here and now, the everyday, it comes very close to a phenomenological approach. Throop and Murphy (2002:197) conclude that Bourdieu's recent work has merely rephrased a number of phenomenological premises in his own idiosyncratic and overly deterministic vocabulary so as to make them sound new, when in reality they are not.

In a fascinating exchange, Bourdieu (2002) responded to the critique of Throop and Murphy at the end of 2001, shortly before he died in January 2002. He argued that he had often declared his indebtedness to phenomenology and that it seemed to him that he had done justice to the work of Husserl, Schutz and a few more, and states that: it was not my intention either to rephrase them [but] to integrate phenomenological analysis into a global approach.

Alfred Schutz (as referred to above) was a sociologist and philosopher who was perhaps most concerned with developing the phenomenological philosophy of Edmund Husserl and his work effectively provides a bridge between phenomenology, sociology and social psychology. Much of his writing focused on the construction of social reality and in particular the ways in which the structure of the lifeworld informs and gives meaning to the lived experience of individuals (Schutz, 1967). Schutz's key argument, neatly summarized by Throop and Murphy (2002), contends that much of what occurs in our daily lives, within a 'horizon of familiarity', is taken for granted.

Schutz argues that social world has a particular meaning and relevance structure for the human beings living, thinking, and acting therein. We navigate our path using a series of common-sense constructs of the reality of daily life. For Schutz, these 'typifications of consciousness' provide a compass to locate ourselves within our 'environment'. They enable us to act in everyday life and to develop a sense of place within the structures of our social reality. For individual actors it is precisely the lifeworld as *experienced* that has consequence for these interactions. Similarly, cultural anthropologists Quinn and Holland (1987) write about the 'presupposed, taken-for-granted models of the world that are widely shared by the members of a society and play an enormous role in their understanding of that world and their behaviour in it'. Such a view, according to Reybold (2002:539), supports both a way of knowing and a way of being. Reybold refers to this as an individual's pragmatic epistemology, the experience of epistemology in everyday life. As Albert Borgman writes in *Technology and the Character of Contemporary Life*, 'concrete, everyday life is always taken for granted ... Therefore it is almost and systematically and universally skipped in philosophical and social analysis. But if the determining pattern of our lives resides and sustains itself primarily in the inconspicuous setting of our daily surroundings and activities, then the decisive force of our time inevitably escapes scrutiny and criticism' (Borgman, 1984:3).

What does all this theory mean in relation to FMD? In Chapter 3 we discussed the moral and ethical issues raised by images of farmers weeping when their stock was culled or burned on pyres. We believe that this distress reflected a severe and often poorly understood disruption to a complex lifescape. The complexity comes from what Tim Ingold (1992) refers to as the mutually constructive relationship between people and place. Put simply, an individual's lifescape, the relationship between people, where *they* are and what *they* do, becomes a product of, and influencing factor upon, *their* environment.

Perhaps the most important question we are trying to raise goes something like this: If one takes Schutz's view that the social world is a horizon of familiarity, and accept that this 'social world' may at least in part be navigated through Bourdieu's emphasis on the non-conscious grounding of social action mediated through the habitus, what then happens when the lifescape breaks down? A sense of reflexivity (which a number of authors have argued is missing from Bourdieu's work) is important in relation to lifescape change. The following extract from a community nurse (which we used in Chapter 5 to show the 'before' and 'after' of a disaster) also serves to illustrate the process of lifescape reflexivity:

> *Going out to Penrith last night my eyes automatically duck down where the pyre smoke used to be. It's not a conscious thing, it just was present for so long and was so depressing, so synonymous with the despair of that time that I look towards it, just as I always check what the hills are like each day. I suppose its become a part of my subconscious landscape.*

More precisely then, how is the lifescape affected when an individual is forced to reflect on his/her agency and context for action in order to reconnect with what Harada refers to as spaces, materials, and the 'social'? We explore this in the second half of the chapter.

Changing relationships, changing lifescapes

As discussed in Chapter 2, even before the FMD epidemic, there were grounds for considering the health and social consequences of a number of crises impacting on UK livestock farmers and their local communities. Indeed, well before the epidemic, the former Minister for Public Health suggested that the 'solution' for such communities would involve, 'rebuilding social capital, civic society, trust, participation, social inclusion and community-based support through strategies which are part of community development' (Jowell, 2000:13). Yet the 'community' in reality comprises a number of diverse groups, with different sets of interests and varying capacities to voice those interests. Furthermore, the 'community' became divided in its responses to the FMD crisis; for example, farmers pressed for footpaths to remain closed in the interest of controlling disease spread, while the tourist industry wished them open in order to attract visitors to the hills. In the previous sections we outlined some of the key theoretical ideas underpinning the lifescape concept. The rest of the chapter focuses on the lifescape shifts experienced by three survivors of the FMD disaster: a livestock auctioneer, a nurse and a teacher.

Lifescape tensions

Nerlich *et al.* (2002) discuss how the FMD disaster exposed many existing tensions, dualities (rural/ urban; farming/non-farming; vegetarian/meat-eater) and stereotypes of different communities. Their study also examined how children talked about FMD, farming practices and stereotypes of town and country on the internet (focusing particularly on one BBC message board – CBBC Newsround website). They relate how children from farming backgrounds felt misunderstood by children from towns who had not lived through the FMD crisis and who, in their view, had no clue about what it is like to live on a farm. Similarly, a DEFRA field officer highlights such tensions, stating that *towns folk just don t understand it, they understand with dogs and cats but they don't see cows, sheep, well not so much sheep but definitely cows, as an extension of that*.

Government management and control of the FMD epidemic also created new divisions within the farming community itself: for example, between those subject to culling and those whose livestock survived. In particular, some high livestock compensation payments received a lot of media coverage, leading to a perception that some farmers had benefited substantially from the loss of their livestock. This placed some farmers and their families under greater stress and social isolation and further threatened community cohesion (although in some instances, community divisions were undoubtedly the manifestation/exaggeration of existing relationships).

> *If you went and asked somebody that's still got their stock and then went and asked somebody that haven't got their stock eh, the government's paid a fortune to them that's lost everything for cleaning out and all this, and the ones 'as kept…managed to keep their stock couldn't even give like free disinfectant'.*

But as another farmer said in poignant understatement:

> *those people that didn't lose their stock, they don't have their memories of that day (*of slaughter*).*

Whilst some rural areas largely escaped infection, all farms were still subject to the severe restrictions on animal movements and sales.

An auctioneer reported: *So many times I have heard customers say to me 'I would have been better off getting the dammed disease'.*

For some rural people, part of making sense of what happened, of telling their FMD story, also captured a sense of old settled relations perceived to have shifted into unsettling terrain:

> *The village has changed quite a lot over the last few months since it was culled in the summer. Some re-stocking of sheep on a small scale but the heart has gone out of some of the village with farmers retiring and changing occupation. Not much sign or noise (good and bad) from farming now. For the first time in 30 odd years I begin to feel a little out of place in the country and also that a slight barrier has gone up between locals and off comers.*

For others, part of this unsettling may relate to different degrees of experience, a revisiting of events that causes them to reflect on previous attitudes.

> *I am writing this diary with truth. The foot and mouth did not traumatise me in any way. I adapted to any restrictions. Mainly my pleasures were disrupted. After the* [project group] *meeting at the George Hotel Penrith I felt GUILTY that I was not upset at the time of the outbreak. At the time I felt sorry for the farmer, but no more than for the coalminers, steelworkers, shipbuilders, factory workers etc etc who have all had troubles. You meet it, deal with it and get on with your lives. Life is a drama and I too love and enjoy life.*

New cultural identities appeared such as 'culled and unculled'; 'front-line and office-based'. The rifts that appeared between farmers whose livestock had been culled and those whose livestock survived were sometimes unspoken: *It's like there are us and there's them.* A farmer notes: *there's been a rift between them that's gone down and them that hasn't.* In some case such divisions were the manifestation or exaggeration of existing relationships, and rifts were sometimes overt and/or violent:

> [Farmers with large compensation payments] *had gone in pubs and been shouting their mouth off and getting people's backs up, he went in the pub but he got three hidings in two or three weeks.*

Those who were isolated as the only farm left 'standing' while others went down around them or who may have been the first to be infected in any given area suffered particular problems: *It maybe got slightly easier for people that went down later. The first ones it really affected them that way.* And another farmer recalls:

> *That first week or so when I was sitting ploughing, and doing jobs sitting on the tractor, just sitting there thinking about it all, there were many a time I had to wipe tears off my eyes, I can tell you. The more you talk about it, the less it hurts. That is what I feel is happening. Some neighbours who you maybe haven't had a lot of contact with over the years are very good and ringing you up at the time every other day to see if you are alright and probably all through summer as well maybe once every two or three weeks just for the crack and to see how you are and it is nice when you think back to think that they did it. There is others who you tried to maybe befriend over the years just a waste of time. I mean obviously down here there was only two farms in this parish where they found foot and mouth and if you go on this land that I have got* (nearby) *all the neighbours there got it and there is a different attitude altogether because they were all in the same boat. So everybody feels a lot easier about it. Around here it was just a bit tense and still is.*

'New FMD identities' also engendered black humour, as this farmer shows describing the signing of the guest book at a wedding in June 2001:

> *So we went to the Church do and then we went back to milk [...] and there were lots of Longtown folk there lost the lot and, when we got up to milk, they said 'You mean you have got stock?', and everybody said how lucky we were to be able to go away in the middle of the wedding back to milk! Everybody had to sign a book to say who they were and they said 'Why don't we put our [FMD] case number next to our name?*[88]

A farmer whose stock survived and who did not leave his farm for ten weeks during the epidemic, speaks of meeting with extended family in Christmas 2001:

> *They didn't get foot and mouth, so the family party was the first meeting we had when nobody had foot and mouth, we could have a good conversation without offending anybody, 'cause there wasn't a foot and mouth victim sitting next to us, and we were just, you know, In some ways makin' fun of them, 'Oh look out here's a foot and mouth victim coming, he's got a new*

Range Rover', or this sort of thing like, and we could say that without actually offending anybody.

Whilst there were certainly tensions in some places, there was also a sense that the events of 2001 also brought community members together. Farmers spoke of the kindness and support they received and how helpful and affirming this was:

You know you have got a lot of friends out there that really do care. You always had somebody running round after you. They were always on the phone I am going down town do you need [anything] could you do this, could you do that. Very kind.

I think at the beginning neighbours were good, especially in villages. Everybody took food to farms. You could see it outside farms there were little packages [and] presents.

Below we consider in more detail how such changes affected the every day 'taken for granted' interaction between people and place and in so doing the nature of lifescape disruption brought about by the FMD disaster.

Disrupted lifescapes

Wilson (2003) has highlighted the culturally specific interactions between place, identity and health, emphasizing the significance of everyday place in people's lives. So when everyday places and routines changed and took on new meaning during FMD, it was as if the 'natural envelope' in which people live out their lives has become defiled and unreliable (Erikson, 1994). Figure 7.1 indicates the complexity of FMD consequences for the nexus of people/place/livelihoods. It highlights how 'taken for granted' family and rural community events were disrupted (events and shows cancelled, weddings postponed) as well as those more obviously related to agriculture (stock culls and mart closures). Combined, such disruption to so many aspects of individual and community lifescapes (work, social and family events) created a deep sense of disorder: *There was no normality, normality had gone.*

In Chapter 3 we pointed out that the social and cultural importance of livestock auctions in Cumbria should not be underestimated. For farmers, the absence of auctions meant loss of social contact and a radical and complex change to trading, with animals going straight to

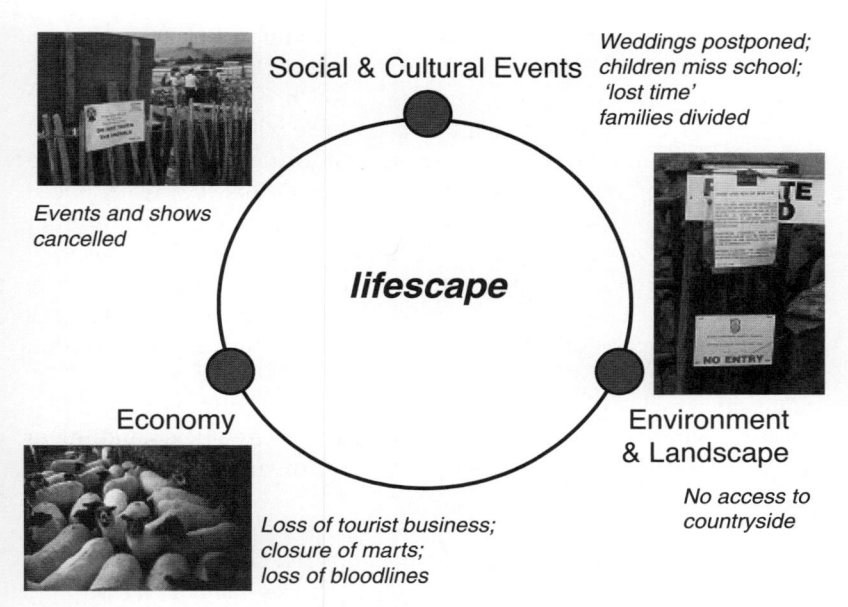

Figure 7.1 Altered lifescapes during FMD

- Social & Cultural Events: Weddings postponed; children miss school; 'lost time' families divided
- Events and shows cancelled
- Economy: Loss of tourist business; closure of marts; loss of bloodlines
- Environment & Landscape: No access to countryside

slaughter often with no agreed final price. Farmers also reported frustration and anger at perceived inefficient, bureaucratic and incomprehensible licencing arrangements. In the longer term, the re-stocking of farms has brought animal health problems, continued stress and for farmers getting used to new livestock, painful reminders of loss. Some report feeling less connected with their work, others say they over-react to problems with their stock which previously they would have managed more easily.

Perhaps the best ways to illustrate all this is to explore in more depth the narratives of three FMD 'survivors'. We do this in order to illustrate how the use of lifescapes can capture 'simultaneously tangible, negotiated and experienced realities of place' (Kearns and Gesler, 1998:4), whilst also highlighting the interconnected nature of the rural economy, the dynamic of individual and community disaster and the complexity of related moral and ethical issues. The interrelations of rural networks have been examined by Donaldson *et al.* (2002) who discuss the effects of the FMD epidemic on rural governance. For us rural governance is something much broader than the act or process of government in rural areas. It refers to the nature of organizations and the involvement of actors and institutions in the development of policy outcomes and the nature of the institutional relationships between statutory and non-statutory agencies and civil society. By analysing people's work, social and personal networks at a very local level, it is possible to show how these inter-relations occur in everyday life. The examples given below illustrate how, firstly, workers have multiple roles, and secondly, how their lives interact with others. The complexity of such interaction meant that they experienced the FMD disaster on many levels. The following extracts illustrate what Howorth (1999) refers to as the temporal and spatial dynamism of lifescapes, reinforcing the interconnections between people & places. The land does not just represent a physical space but rather, represents the interconnected physical, symbolic, spiritual and social aspects of (rural farming) culture.

The Livestock Auctioneer. In his retrospective telling of his experiences of living and working with the consequences of the handling of the FMD disaster, Eddie graphically recounts disruption to family and community (below and Figure 7.2):

> *I'm 37 years of age. I was brought up in the Cockermouth area, in a hill farming area. I use to go to Ulverston auctions as a child, and with my*

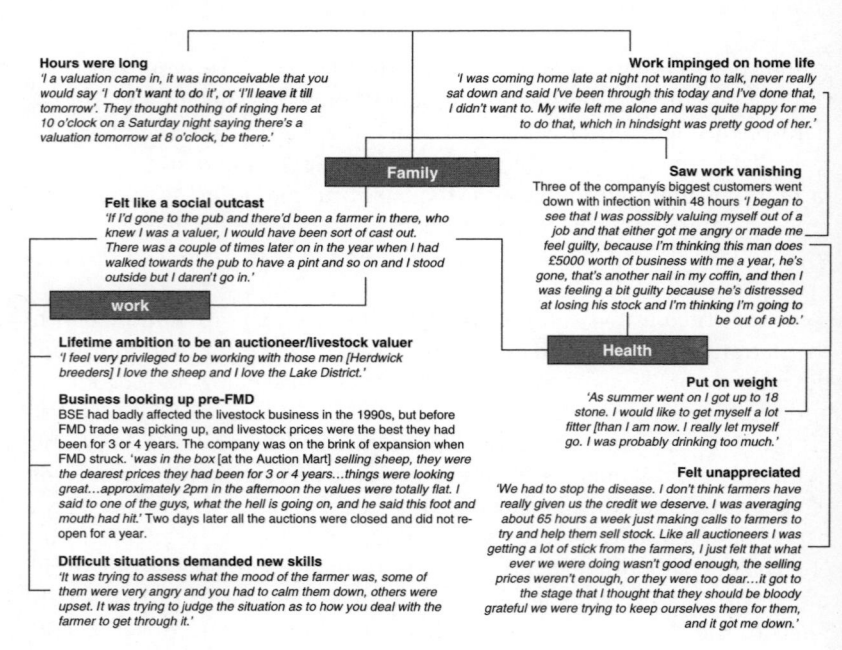

Hours were long
'I a valuation came in, it was inconceivable that you would say 'I don't want to do it', or 'I'll leave it till tomorrow'. They thought nothing of ringing here at 10 o'clock on a Saturday night saying there's a valuation tomorrow at 8 o'clock, be there.'

Work impinged on home life
'I was coming home late at night not wanting to talk, never really sat down and said I've been through this today and I've done that, I didn't want to. My wife left me alone and was quite happy for me to do that, which in hindsight was pretty good of her.'

Family

Felt like a social outcast
'If I'd gone to the pub and there'd been a farmer in there, who knew I was a valuer, I would have been sort of cast out. There was a couple of times later on in the year when I had walked towards the pub to have a pint and so on and I stood outside but I daren't go in.'

Saw work vanishing
Three of the company's biggest customers went down with infection within 48 hours 'I began to see that I was possibly valuing myself out of a job and that either got me angry or made me feel guilty, because I'm thinking this man does £5000 worth of business with me a year, he's gone, that's another nail in my coffin, and then I was feeling a bit guilty because he's distressed at losing his stock and I'm thinking I'm going to be out of a job.'

work

Lifetime ambition to be an auctioneer/livestock valuer
'I feel very privileged to be working with those men [Herdwick breeders] I love the sheep and I love the Lake District.'

Health

Put on weight
'As summer went on I got up to 18 stone. I would like to get myself a lot fitter [than I am now. I really let myself go. I was probably drinking too much.'

Business looking up pre-FMD
BSE had badly affected the livestock business in the 1990s, but before FMD trade was picking up, and livestock prices were the best they had been for 3 or 4 years. The company was on the brink of expansion when FMD struck. 'was in the box [at the Auction Mart] selling sheep, they were the dearest prices they had been for 3 or 4 years...things were looking great...approximately 2pm in the afternoon the values were totally flat. I said to one of the guys, what the hell is going on, and he said this foot and mouth had hit.' Two days later all the auctions were closed and did not re-open for a year.

Felt unappreciated
'We had to stop the disease. I don't think farmers have really given us the credit we deserve. I was averaging about 65 hours a week just making calls to farmers to try and help them sell stock. Like all auctioneers I was getting a lot of stick from the farmers, I just felt that what ever we were doing wasn't good enough, the selling prices weren't enough, or they were too dear...it got to the stage that I thought that they should be bloody grateful we were trying to keep ourselves there for them, and it got me down.'

Difficult situations demanded new skills
'It was trying to assess what the mood of the farmer was, some of them were very angry and you had to calm them down, others were upset. It was trying to judge the situation as to how you deal with the farmer to get through it.'

Figure 7.2 Disrupted Lifescape: Eddie

father's and grandfather's grace, I use to sit up alongside an auctioneer so it's always been in the back of my mind...Normally you go out on a farm and you have a laugh and a joke, you value the stock for them and you do your job professionally, this was different, this was trying to keep the farmers upright, trying to stop them from bursting into tears, or to control it if they did burst into tears. I had times when I had farmers in tears, vets in tears, and slaughter men in tears, and that's bloody hard to know what to do.

There was no time with the family, and that made me feel guilty because P was seven months pregnant at the time, with a two year old daughter as well. She was tired and struggling and I was coming home late at night not wanting to talk... social life was very difficult for me.

I saw some cattle that were quite distressed and I thought well here we are, I've seen my very first case of foot and mouth disease...barely able to walk, frothing and snotty nosed. Some of them were in a bad sort of state and their tongues were starting to drop off. The vet said to me, 'there are two things that have to happen immediately here. First thing we have to get the army in, we must get the army in, we can't cope with it. And secondly, these animals have to be slaughtered and removed off the premises, treated, you know, whichever way, in 24 hours'. And that was on the Monday when I did the valuation. The farmer first noticed that he had foot and mouth the previous Wednesday, and it was the following Sunday before they were put on a pyre and burned up. Basically for them, the 12 days those animals shed viruses, and the vet said,' if we don't do something better than this, this virus is going to spread through the county'. The vets on the ground knew that was the case. But the people in London didn't, or they didn't appreciate, or they didn't want to know, what ever way it was, they let us down.

There is also an ongoing sense of frustration at having to deal with post FMD regulations that Eddie perceives to be 'unworkable.'

The rules and regulations which DEFRA are imposing on markets are just ridiculous. They're out to make sure that we don't operate. We want the markets, our farmers desperately want the markets. The only people that don't want the markets are DEFRA and they're blaming the sites nothing else but the sites, it's just ridiculous.

He writes that whilst DEFRA view animal movements:

...as the root cause of the problem, they consistently ignore how the disease got into the country. We've been so badly let down. And to say to

the farmers, 'no I'm sorry, but you're not going to be allowed to move your animals as you used to', is not good enough, that is not good enough. All they're doing, is basically, putting such pressure on the farmers, to make their businesses unworkable as well. So I'm very concerned about the future of the business.[89]

The Nurse (Carole) works at a large GP practice and is also part of a wide extended family of farmers. She has two teenage children. Her clinical work was important to her during FMD and she describes *my life at work* as being *minimally affected* by FMD compared with her life at home. She says this made her hesitate to join the research panel; however she wrote in her first diary:

After a few days thought I decided it would be good for me to do this diary thing. It may help me come to terms with what has happened, and in years to come I may regret not writing this. FMD has brought about enough regrets without adding to the regret list.

FMD made her rethink her career and think about further training after the cull on the family farm: *until FMD never had the time. Sometimes I feel this is positive result of FMD.* But in the end she decided against it: *no concentration. I used to read stuff and understand it – why can't I do this any more?* Since theirs was a big farm where stock was slaughtered in the buildings and there was a large pyre, they had 'strangers', DEFRA officials and cleaning and disinfection teams, on their farm for months and she resented what she saw as the continual invasion of privacy. She talked of a dividing line between her work at the surgery and her experience of the cull, how she maintained a professional distance and did not talk about her own experience when patients would attend with problems: *I would hate to think that anybody came to see me and couldn't have told me things because they would distress me.* There were times when she used strategies for dealing with *difficult matters*, things which could cause her distress or discomfort. *There are things that people have come with (...) I've sort of put them in another direction.*

She is actively involved with community activities; has a busy social life and wide circle of friends. Yet she expresses deep disillusionment and loss of trust in government:

They have no credibility at all in my eyes now, because I know what they used to say to the Press in London and what was actually going on in reality up here, and you know it's, it's very very sad that you've lost, it's lost its mystique of always being there in a crisis or coming up sound and what have you, when it let you down badly.

This has led her to view 'official' inquires and reports with suspicion. For example, in 2002 she attended a public meeting of the Cumbria FMD Inquiry and writes that she *'felt that it was a strange experience. Felt we had heard all this stuff before SICK OF FMD! Pleased when meeting was over. Had gone out of duty only. Feel all the inquiries will have little impact or the impact they SHOULD!.*

Figure 7.3 demonstrates how the immediate effects and residual 'fallout' from FMD impacted on Carole's lifescape, from her relationships with immediate and extended family (her children experienced disturbed sleep and nightmares and her own health deteriorated) through to issues of trust in governance and community identity. Significantly, the everyday place of the farm, a central component of her lifescape, became associated with death, the family stock was culled within farm buildings and the 'family trauma' was on 'public view', as the pyre was situated close to a main road. Carole's post-FMD lifescape illustrates fractures across what Ingold (1992) refers to as mutually constitutive relationship between

Family

Extended family
Members of extended family also farmers, FMD on their farm led directly to a livestock cull on another farm.

Identity
Pride in farming has been replaced by the 'stigma of FMD victim.'

Home
Home no longer place of safety – 'invasion of officials' (daily from early Spring to November 2001). 'That was the first time since March that there hadn't been a stranger on the farm'

work

On-Farm
Fragility of recovery 'Very upset...totally over-reacted' to health problems in new stock. The return to 'normality' is slow, 'the pyre field is still not back to its original use.'

Off-Farm
Health professional – trying to maintain professional identity and keep work and home separate

Cull location
Their cull took place in farm buildings and the pyre was located in a field between the farm and the road. All these events were 'on public view' – she felt stigmatised.

'Community'

Traditions
She feels deeply embedded in local community. 'Participates in local shows but now no livestock. Irritated by presence of 'stress organisations' at shows, 'all this negativity'.

Shared Experience
FMD recurring topic of conversation with friends in 2002 'almost competitive'.

Cynical & Distrustful
'I question and I am cynical', She now feels cynical and mistrustful towards 'authority' where she was previously trusting.

Trauma – Recovery
Children exhibit enduring symptoms (including sleep disturbance 15 months after the cull). **Husband** distraught at time of cull, wants to return to normality and consign FMD to the past

Health

Usually Energetic
'I can't concentrate, I used to read stuff and understand it.'

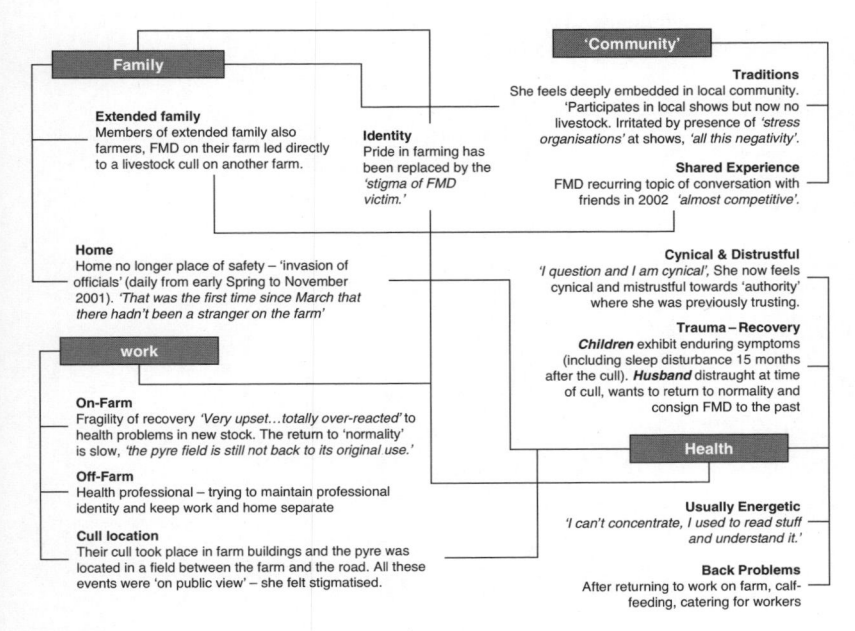

Back Problems
After returning to work on farm, calf-feeding, catering for workers

Figure 7.3 Disrupted Lifescape: Carole

148 *Animal Disease and Human Trauma*

people and place. Ingold highlights the way in which by moving in familiar fields, forests and paths, creating memories and stories about places, people create a sense of self and belonging and an intimate knowledge of their environment.

The Teacher (Jean) lives with her husband in a rural area and drives 12 miles along narrow roads to work. There were many pyres along this route, and she recalls how she drove down here one beautiful spring day, and counted twelve pyres on my way down to (work), twelve! Driving in such a sparsely populated area she saw the same people every day. Particularly:

> *Early in the morning you see a lot of people in tractors and you, you know you get to the point where you recognise people and I've been doing it five years so I recognise everything.*

Although not involved in farming herself she was deeply emotionally affected by what she saw during 2001, and sometimes she was unavoidably caught up in distressing circumstances (below and Figure 7.4):

> *And I was stopped like that several times, different farms, and just had to sit there while they got on with stuff. The worst one was later when they*

Family

Husbands work connected with FMD. Her son-in-law also worked on FMD 'he used to come in coated with (ash).'

Daughters
'1 separated post-FMD, 1 expecting baby and soon to be married. Both have chronic health problems. Both their partners worked on FMD disposal. Before he went on that he was sent for a medical. I said 'are they sending you for a medical afterwards?' He said 'I don't think so.'

work

Lasting images of FMD
'If I see a man in a white suit, especially if it's got a hood up, I get a cold shiver…the hair stands up on the back of my neck.'

Parents needed to confide in her
'The (parents) needed to tell somebody, they needed to say what they'd seen…and you've got to listen…it can be quite traumatic sometimes.'

Guilt and worry about travelling to work
'It made it very difficult, I felt so bad driving from here to…get to work. I just didn't want to be responsible for spreading it around, but I couldn't stay off school.'

Daily commute to work
'I watched them all burn…everyday I went past they burned for about four weeks….it was like driving in thick fog in places… I drove down here one beautiful spring day and counted 12 pyres.'

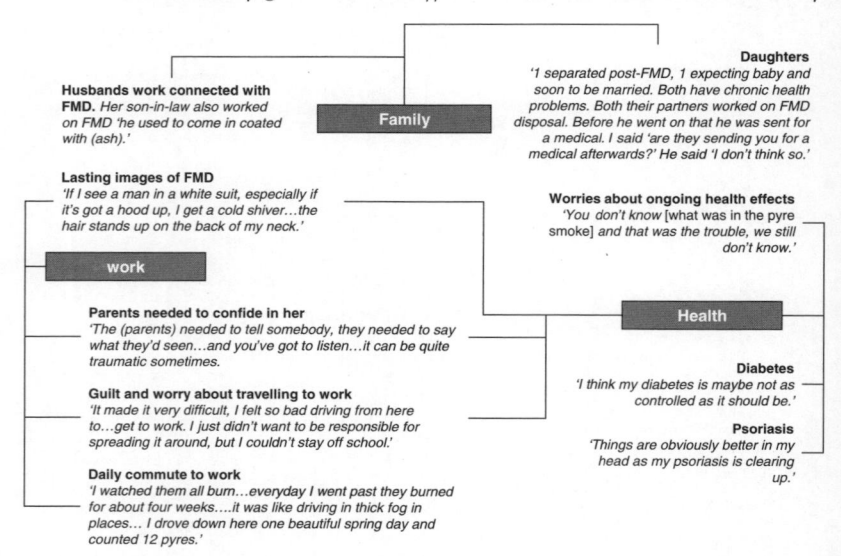

Health

Worries about ongoing health effects
'You don't know [what was in the pyre smoke] and that was the trouble, we still don't know.'

Diabetes
'I think my diabetes is maybe not as controlled as it should be.'

Psoriasis
'Things are obviously better in my head as my psoriasis is clearing up.'

Figure 7.4 Disrupted Lifescape: Jean

had killed them all and then had to pick them up and take them to the different places, because that was just, the stench [...] some of those animals were left ten days before they were picked up.

At school she could not avoid the FMD narrative of parents who had been affected because *they needed to tell somebody [...] and you've got to listen*. She also worried for the children in her care, and was particularly concerned about the effects of the surrounding pyres. She notes how:

There was a few, good few days where, when it was playtime we just didn't take the children outside because, of the smell as much as anything. I mean it was making them gag.

Her daughter's partner worked on the disposal of carcasses, work which he found very distressing. On one occasion *he spent three days once just chopping cows in half with a bucket on his machine*. She worried for long-term health impacts of this work and advised him to watch *what you're doing it's not worth getting a week's salary now to end up with, a severe illness in your early forties, or late thirties*.

Jean has also developed a loss of trust in governance and a lingering sense of injustice, and feels that she will never find out 'the truth' about FMD.

I find that really hard that they can hide things for 30 years, I think that's dreadful. I want that changed. It might be supposedly for our own benefit. Not sure that that's going to wash, might wash if you're a politician with things to hide, but not if you're a member of the public who wants to know.

These three 'stories' highlight disruption to lifescape: place, occupation, family, education and culture and an articulation of the view that Cumbria was 'too far away to matter'. New cultural identities appear. Above all there is the strong sense (well known in disaster studies), that unless you were there, it is impossible to explain, and this gap in comprehension/empathy is a continuing sense of frustration. This creation of 'insiders and outsiders', a kind of collective identity, is discussed by Erikson (1994) from his work in many different disaster settings, but here it is summed up eloquently by a Cumbrian rural vicar:

I try to explain what it was like to be here through last year but am still unable to articulate thoughts and feelings accurately. It is so hard to try to enable people who weren't here to get a real grasp of it. The fear, the anger, the frustration.

Summary

In our work on FMD, we focus on lifescapes because of the emphasis on the consubstantiality of people, place and livelihoods, a dynamic which was severely disrupted by the events of 2001. But in using lifescape as a conceptual framework for understanding the consequences of FMD, we also recognize that there are many other ways of thinking about the relationship between people and place. At the start of this chapter we attempted to locate lifescapes within a broad body of literature, including sociology, geography and phenomenology.

Everyday places changed dramatically during FMD, some of these have altered in the longer term or permanently, for example in the case of disposal sites. While the pre-FMD 'lifescape' undoubtedly contained hardships and continual changes, the alterations brought about by the epidemic affected peoples' sense of identity, self-esteem and well-being. Journeys to work took people past pyres, mounds of dead livestock and the logistical traffic that went with the culls. The sense of FMD identity is mentioned in reminders such as tarmac strips where disinfectant mats were laid on roads and how only those who lived through it would know what these represented. Memory landmarks were forever changed in this cognitive landscape, overlaid with mainly bad associations. The increasing sense of social isolation among livestock farmers as a result of FMD (alongside other agricultural 'shocks' and significant recent policy changes) has undermined farming as a way of life and the cultural identities of those working in the sector.

We have used the concept of lifescapes to try to articulate that which hybrid and collective and conveys a 'way of life'. We argue that communities in Cumbria are at once defined by their lifescapes and define their lifescapes, which in turn were changed because of a catastrophic series of events. The 2001 FMD epidemic created deep fissures in the lifescapes of Cumbria, so that much of the taken-for-granted world, identity and sense of meaning changed. As Minh-ha (1994:15) indicates 'every movement between here and there bears with it a movement within here and within there.' The consequences of this process are likely to be long lasting.

8
Reconfiguring Disasters

There is general agreement that a disaster is some event or circumstance outwith normal (i.e. situated) expectations of life and that coping with its consequences requires skills and/or resources beyond those available in the affected community. Disasters previously classified as natural are today considered, to an ever increasing degree, to be human induced (Weisæth et al., 2002), and in this book we have likewise argued for a more nuanced understanding of disaster; what Erikson calls 'a new species of trouble'. Disasters are political events, both in terms of government and governance.

The 2001 FMD crisis is commonly understood to have been an animal problem, an economic industrial crisis, which was resolved after eradication. But if you use a different lens, in our case a longitudinal ethnographic one, it can be seen as a human tragedy, as well as an animal one. By listening to individual voices, we better understand how people align their personal experiences and the wider disaster, what Harada terms the individualization of the disaster.

Returning to Aberfan more than 30 years after the disaster and talking with survivors, accessing public records previously embargoed under the 30 year role, McLean and Johnes (2000) were able to show how the distress and anger of disaster victims may be hidden or misunderstood at the time, but is in effect a normal and justified response to events outside of their control. In a diary based study, it can be seen that life after the FMD crisis was accompanied by distress, feelings of bereavement, fear of a new disaster, loss of trust in authority and systems of control, and the undermining of the value of local knowledge. Distress was experienced across diverse groups well beyond the farming community. Such distress remained largely invisible to the range of 'official' inquiries into the disaster. Moreover, a 'disaster' in the more conventional sense, such as the 2004 Tsunami or the sudden and horrific passenger flight bombing over Lockerbie in Scotland tends to prompt or sanction pro-active and innovative approaches on the part of relief service providers, who feel able to dispense with 'peacetime' or normal rules. This frequently is not the case with more insidious enduring disasters such as the poisoning of the Ojibwa land at Grassy Narrows or the wholesale slaughter of 'life's work' in Cumbria.

The scale and reach of the FMD disaster both within and beyond farming was greater than has been understood. The Lancaster study showed the profoundly interdependent nature of human-animal, of farming and related industries, and of the rural economy itself. It also made visible an important feature of traumatic experience: that the longer the exposure (in the 2001 FMD crisis eight months of outbreaks and 12 months of severe restrictions), the more lasting and complex the effects.

We have highlighted the trauma and emotional pain of implementing culling policies to deal with animal disease. In particular, how the deep bond that exists between farmers and their livestock has been largely ignored or overlooked in devising responses to animal disease. As Harada notes, our individual and social relations are always constructed and transformed by space and materials. When these things change, so also does our sense of identity and connection. Farming is much more than a job or even a way of life, for many farmers their farm and their livestock represent a fundamental element of their identity. As Hall *et al.* (2004) note, appreciating the nature of humans' attachment to their animals and the meaning of this relationship in different sociocultural and occupational groups has enormous practical implications for disaster management and a critical element in promoting the resilience of individuals and communities. This approach necessitates a more flexible, innovative approach to rural health outreach work in times of crisis. In Chapter 5 we discussed an approach developed during an outbreak of Ovine Johne's Disease in Australia (Hood & Seedsman, 2004) to support farmers whose livestock was about to be culled which involved the offer of 'counselling' support from their *local* vet. This is in line with a non-pathologizing approach to traumatic experience: people are not sick but do need help, and furthermore it is best if this help comes from trusted actors within their network.

Again psychological trauma encompasses both the events, and how those events are experienced by individuals and communities who may be trapped and unable to take control over their place relative to events. It has particular characteristics in disaster settings and Erikson highlights the feelings of distrust, usually concerning local and national institutions, people who have gone through a disaster often develop. It is vitally important that wherever possible, communities are actively involved in making decisions about the management of disasters. For communities living close to the FMD animal disposal sites, this either did not happen or was merely 'lip-service' consultation. Likewise, local knowledge and expertise was largely ignored in favour of a centralized approach to managing the disaster. We heard ample evidence in discussion groups and interviews of the frustration and despair felt by people who saw a situation spiralling out of control while they were powerless to act. This was most strongly felt by those who believed that they might have been able to use their own expertise to help alleviate the situation had they been given the opportunity. More specifically, a gap opened up between different kinds of knowledge: proximal knowledge derived from local experience, and centralized or distal knowledge. For example organizational directives were perceived to be stripped of context, unable to adapt to what was happening 'on the ground' and unable to mobilize stocks of local expertise. This manifested itself most devastatingly with the contiguous cull policy and the way in which it ignored local knowledge and agricultural 'common sense' (Bickerstaff et al., 2006).

We have also highlighted the plight of frontline workers, too often considered to be infinitely resourceful and impervious to traumatic experience. Yet the industrial-scale slaughter of FMD 2001 was deeply traumatic both for those on the receiving end and, in a variety of ways overlooked by official inquiries, for many of those frontline workers heavily involved in carrying out the slaughter and disposal. While many of these workers were local people whose livelihood had been severely curtailed by the FMD control strategies and who had practical knowledge of handling livestock, others were conscripted or seconded from a huge range of different occupations and parts of the world with very little prior training or preparation. We have outlined the need for debriefing time and peer support for frontline workers. This is more likely to be successful if it is 'informal' and comes from within the 'social capital' of the team or organization, rather than from an outside team of expert counsellors. Frontline workers amassed huge expertise during FMD. We argue that this is still not being sufficiently recognized or recorded. It is also important that agencies seconding front-line workers should record of skills and expertise acquired and ensure ways to call on this in future (and to also provide the necessary pre-event training and preparation).

In Chapter 7 we explored the theoretical frame and history of the lifescapes concept, identifying that whilst it is drawn from phenomenology, it also has resonance in geography and sociology. In particular, the interconnectedness embodied by 'lifescapes' intersects with health geography and social anthropology, such as the work of Wilson and Ingold. The narratives used to convey the complexity of lifescapes, that is to say virtually all the empirical material in this book, communicate the complexity of disasters. They are irreducible. Each reading reveals new understandings and meaning and repays repeated visits.

In any disaster situation affected communities need to have the chance to regain some control of their own lives, personal spaces and collective space; they must not be seen as 'the problem' rather that the solution. They need to be at the centre of decisions and actions that affect them rather than 'othered' and managed. Participatory approaches are not easy, things get messy and disaster managers and regeneration agencies tend to like things to be clear and ordered. But disasters are not like that. Of course, it is often very difficult to define the parameters of a 'community', and particular disasters create particular kinds of communities. Nevertheless, such communities can only be understood by spending time alongside people affected and by listening. It can be time consuming to commence and sustain such engagement. In the heat of a disaster situation it may be tempting to opt for quicker, less troublesome solutions, particularly when the community response appears 'illogical' to those 'outside' the disaster, as graphically illustrated by the inhabitants of New Orleans, Buffalo Creek or Aberfan. This would be a mistake. Lack of participation and genuine engagement between disaster management and the affected community is likely significantly to increase recovery time and may exacerbate the likelihood of longer-term social dislocation, trauma and associated mental health problems.

When we started this work we made a promise that we would bring the story of what happened in Cumbria in 2001 to as wide an audience as possible. We hope that in writing this book we have gone some way to do justice to the commitment, honesty and bravery of the panel members who worked with us on this project. That an FMD epidemic of the scale of 2001 could happen again in a developed country is widely accepted and is a deeply worrying prospect.[90] It is to be hoped that contingency plans are evolving along with enhanced understanding of the human, as well as financial cost of the disaster. Such plans, we would argue, must redefine what counts as a disaster. This necessarily involves much closer working between the people affected, the informal and voluntary organizations that tend to spring up or step up during disaster situations, and the formal, visible, state services. But above all else it requires that people affected by disaster speak, and that we allow ourselves to listen.

Appendix I
The Compulsory Cull

Press release issued by MAFF Carlisle on 27th March 2001. Foot And Mouth Disease: New Measure To Help Prevent Spread Into Lake District.

'MAFF today announced that culling of sheep within three kilometres of confirmed foot and mouth outbreaks in Cumbria will begin today Wednesday 28th March. Detailed proposals for the cull have been drawn up by MAFF working in close consultation with the National Farmers' Union. The measure is aimed at helping to ensure that the disease does not spread into the Lake District. As a matter of priority, the cull will start on the outer most confirmed cases in the firebreak area which follows a 'U' shape with the base of the 'U' around Penrith. The area includes the three kilometre zones around infected premises on the Solway. The scheme is compulsory. Sheep will be valued and compensation paid at the full rate. Local auctioneers have been contacting farmers within the 'firebreak' area since last Friday and some valuations have already taken place. Sheep will either be taken live from the holding to a central point for slaughter followed by burial or burning, or they will be killed on site and then buried or burned. The vast majority will be removed to a central site. About 6,000 sheep will be taken to Great Orton airfield for slaughter today. Additional sheep will be taken to Carlisle abattoir for slaughter and subsequent burial. Farmers who feel that they have a strong case for exempting their flocks from this compulsory cull will be able to appeal to MAFF. Details of the appeal process will be made available as soon as possible.'

The Deputy Chief Veterinary Officer (DEFRA) also issued a note (which was subsequently circulated to farmers by the NFU on the 7th April 2001) outlining the slaughter policy with regards to contiguous premises; in particular, the note indicated the DEFRA position if a farmer objected to the contagious cull:

Review: If, in spite of the explanation of the strategy, the farmer objects to having his stock taken, a rapid assessment should be carried out by a veterinarian in order to: (a) confirm whether or not the premises is in fact contiguous. There are geographical issues (such as a significant area of woodland being in the way) which may affect this judgement; and (b) confirm the veterinary assessment that the livestock on the contiguous farm concerned appear in any way to have been exposed to infection. The fact that animals have been housed does not of itself mean that they have not been exposed to infection. This is against the background of the general judgement, in all the circumstances of this epidemic, that susceptible animals on contiguous premises will have been exposed to infection. It should be emphasised that the purpose of the review process is to establish the application of the general policy in the particular case. In all cases of farmer objection, a careful record should be kept of the action taken, and the grounds of objection, and the grounds on which any objection was overruled or, as the case may be, allowed. The final decision in cases where the farmer persists in his objections should be taken by the Divisional Veterinary Manager consulting if necessary with Senior Veterinary Management.

Enforcement: Enforcement should be handled positively and firmly but with understanding and sensitivity. If a farmer has had his case reviewed as above and the decision remains that his stock should be taken, further attempts should be made to persuade him to comply in the interests of all other farmers in the region. If, ultimately, persuasion does not succeed, it should be clear that Ministers are entitled to secure the implementation of their decision. This is because of their view that the animals should be slaughtered because it appears that those animals have been in any way exposed to the infection of foot and mouth disease. Police Officers may enter onto land as a result of the combination of the powers under the Animal Health Act 1981 and their common law powers, in order to secure enforcement.

Notes

Introduction

1 All the extracts directly quoted in this book (i.e. the ones in italic) are taken from respondents (unless otherwise indicated). This study was undertaken by the Institute for Health Research, Lancaster University which received funding from the Department of Health. The views expressed in the publication are those of the authors and not necessarily those of the Department of Health.

Chapter 1   Disasters

2 According to CRED, the number of people affected by disasters over the three decades from 1960 to 1990 has increased almost nine fold from an average of 23 million people affected each year during the 1960s to 206 million people affected each year in the 1990s, an increase of almost 800%.

3 The UK Government no longer uses the term 'disaster' to define an incident, and has for the past few years used the term 'emergency'. The 'official' definition of an 'emergency' can be found in Chapter 1 para 14 of *"Emergency Preparedness"*, the official guidance to accompany the Civil Contingencies Act, 2004. The electronic version can be found at http://www.ukresilience.info/preparedness/ccact/eppdfs.aspx. An emergency is defined as: 'An event or situation which threatens serious damage to human welfare in a place in the UK, the environment of a place in the UK, or war or terrorism which threatens serious damage to the security of the UK.'

4 On 26 April 1986, during a test of the ability of the Chernobyl nuclear power plant (located in the former Soviet Union near Pripyat in Ukraine) to generate power while undergoing an unplanned shutdown, safety systems were turned off, leading to an explosion and nuclear fire that burned for ten days. The explosion transported vast amounts of radioactive material into the atmosphere, much of which was subsequently deposited not only in the immediate vicinity of the power plant in Ukraine, Russia and Belarus, but also led to radioactive material being deposited over Eastern Europe, Western Europe, Northern Europe, and eastern North America. The accident caused severe social and economic disruption and had significant environmental and health impacts. The accident caused the deaths of 30 power plant employees and fireman within a few days, mainly from acute radiation sickness. The WHO (2006) have attributed 56 direct deaths (47 accident workers, and nine children with thyroid cancer), and estimated that there may be 4,000 extra deaths due to cancer among the approximately 6.6 million most highly exposed individuals. The report also cited 4,000 thyroid cancer cases in Belarus, Russia and the Ukraine among children and adolescents for the period 1992–2002.

158 Notes

5 In one of the world's worst industrial disasters, a failure in a holding tank at the Union Carbide plant in the Indian city of Bhopal (in the state of Madhya Pradesh) led to the release of 40 tonnes of methyl isocyanate (MIC) gas. This led to the immediate deaths of nearly 3,000 people and ultimately caused the deaths of between 15,000 to 22,000 people. Amongst survivors of the disaster, people have developed breathing difficulties, various cancers, serious birth-defects, blindness, gynaecological complications and other problems.
6 The concept of the 'other' has been used to explain gender difference, colonialism, security, racialization, labour market position, the stigmatization of the disabled, homeless and mentally ill people, and other forms of socio-spatial exclusion. According to de Beauvoir (1972), 'otherness is a fundamental category of human thought ... as primordial as consciousness itself'. Shurmer-Smith & Hannam (1994) have suggested that as soon as we start to think about people who are not ourselves, we lapse into the language of 'othering', a position where the other is secondary. Drawing borders around territory to produce 'us' and 'them' does not simply reflect divisions, but also helps to create difference. As Augé (1998) remarks, it is intolerance and fear of the other that itself creates and structures *othernesses*.
7 See http://www.massobs.org.uk/index.htm
8 See www.lancs.ac.uk/fass/projects/footandmouth. See also http://www.esds.ac.uk/qualidata/support/q5407.asp

Chapter 2   Global to Local: The Case of Foot and Mouth Disease in Cumbria

9 Figure relates only to infected premises, not to those which were culled because of dangerous contact or contiguous premises.
10 George Fleming, writing in 1871 after the 1865–66 outbreak of rinderpest, states that 'the losses from only two exotic bovine maladies ("contagious pleuropneumonia" and the so-called "foot and mouth disease") have been estimated to amount ...to 5,549,780 head, roughly valued at £83,616,854' (cited in Reynolds & Tansey, 2003).
11 MAFF was later subsumed into the new Department for Environment, Food and Rural Affairs (DEFRA) after the general election in June 2001.
12 Similar descriptions were commonplace in much of the UK media.
13 The National Audit Office (2002:37) note that the 1999 Drummond report on preparedness across the State Veterinary Service found considerable variations in the Service's readiness to deal with outbreaks of exotic notifiable diseases, including foot and mouth. Existing contingency plans in many areas had not been updated because of other priorities and limited staff resources. The Drummond report expressed concern that a rapid spread of foot and mouth disease could quickly overwhelm the State Veterinary Service's resources, particularly if a number of separate outbreaks occurred at the same time.
14 NFU Bulletin, 23rd January 2008.
15 It is outside the scope of this book to address the issue of vaccination, but as the National Audit Office (2002) indicate, at the height of the outbreak the Government accepted that there might be a case for a limited emer-

gency vaccination programme and the Department began to draw up plans to vaccinate cattle in Cumbria and Dumfries and Galloway and possibly Devon. The necessary support of farmers, veterinarians, retailers and food manufacturers was not forthcoming, however, and vaccination did not go ahead. Moreover, according to the Cumbria Inquiry (2002), for a country such as Britain, which imports food from all over the world, the serotype specificity of the present generation of vaccines is a significant limitation. Whilst a vaccination strategy could therefore be based on the serotype of greatest perceived risk to the UK, this would provide no guarantee of protection since illegal imports of food (and different serotypes) could come from almost anywhere in the world. The development of improved vaccines with a broader specificity appears to scientifically feasible, and as they become available the dilemma of vaccine selection will be resolved. However, it is also important to note that under the present EU policies (Directives 85/51/EEC and 90/423/EEC), vaccination against FMD is not permitted except as an 'emergency vaccination' to deal with a FMD outbreak.

16 This is, of course, a problem not confined to Cumbria. For example, a recent analysis of hill farm incomes in Derbyshire revealed a 75% drop over ten years (Peak District Rural Deprivation Forum 2004).
17 These areas were in the main uninfected, but movement and trading and tourism restrictions meant a loss of income, coupled with increased costs for feeding extra stock and disinfection.
18 Whilst the last reported case of FMD was on 30th September 2001, there were still restrictions for livestock movements and footpaths closed over affected farms in early 2002.
19 Burning on pyres (which was a hugely unpopular policy stopped in early May 2001) and burial in other existing landfill sites were also used for carcass disposal while older and infected cattle were taken to a rendering plant at Widnes.
20 The Anderson Inquiry report states that there were 2,030 infected premises in UK (including 4 in Northern Ireland) and that pre-emptive culling was carried out on a further 8,131 premises.
21 The total number of animals culled was officially reckoned to be 6,456,000. However this does not include the numbers of young lambs and calves which were not included in the final figures, estimates vary but it is thought that at least a further 1.2 million young lambs were also killed. The official total of 6,456,000 can be subdivided as follows: sheep: 3,428,000 (of which 968,000 at confirmed infected premises) and a further 1,821,000 killed on welfare grounds; cattle: 592,000 (of which 301,000 at confirmed infected premises) and a further 166,000 killed on welfare grounds and pigs: 143,000 (of which 22,000 at confirmed infected premises) + a further 306,000 killed on welfare grounds.
22 The National Audit Office (2002) state that the direct cost to the public sector is estimated at over £3 billion and the cost to the private sector is estimated at over £5 billion.
23 The main categories of restriction were covered by the provisions for *Infected Area* and *Controlled Area* declarations and by the use of *Forms A, C* and *D*. (*Form B* and *Form E* were used to lift the effects of *Form A* and *Form D*, respectively). When disease was first suspected a *Form A* was served on the premises,

declaring it to be an *infected place*. A 'Form A' notice placed severe restrictions on movements to and from the premises and prohibited any animal, person or object entering or leaving the premises without permission, and a series of measures was required to establish biosecure isolation of the premises. After clinical examination by a DEFRA vet, who confirmed the suspicion that disease was present, a *Form C* was signed. This prohibited movements of animals within an 8 km radius of the place of the suspected outbreak or on any road, rail or motorway passing within 3 km, except under licence. On formal confirmation of the disease by laboratory test or clinical signs, an *Infected Area* was declared, based on a minimum distance of 10 km radius. Livestock movements were banned and rigorous biosecurity measures put in place. All premises within a 3 km *protection zone* area were placed under *Form D* restrictions. An *Infected Area* designation could be lifted after clinical examination of cattle and pigs and blood testing of sheep within the *protection zone* had proved negative. However, this was not allowed until 30 days after the preliminary cleansing and disinfecting of the infected premises. A *Form D* was given to premises, both within and outside an *Infected Area*, if there were reasonable grounds for suspecting that an animal has been exposed to infection. It imposed a ban on animal movements, except under licence, and introduced a requirement for rigorous biosecurity measures to isolate the premises (from Cumbria Inquiry Report, 2002:28).

24 The Ministry of Agriculture, Fisheries and Food, the government agency initially responsible for dealing with the disaster was later 'realigned' in Spring 2001 as the Department for Environment, Food and Rural Affairs.

25 The 30 month cattle rule: cattle over 30 months were not meant to be buried because of the existing regulations controlling spread of Bovine Spongiform Encephalopathy (BSE). Also relevant are the Drummond recommendations of slaughter within 24 hours and disposal within 48 hours.

26 According to the Cumbria Inquiry (2002), when FMD was discovered to have been at Longtown, the Market provided a full list of contacts to MAFF, including details of sheep purchasers on 15 February and 22 February and of market staff who worked or lived on agricultural holdings. However, of the 36 personnel in this category, 34 had still not been contacted 6 weeks later. Similarly one of the buyers who had taken stock to Derbyshire was not contacted for 18 weeks. Copies of the Market's records were requested by MAFF a month after they were first supplied, as the originals had apparently been lost.

27 Aside from the now notorious pronouncement to cull all animals within 3 km of an infected premises, the measures also made provision for: intensifying surveillance controls within 3 km of infected premises; slaughtering sheep in 3 km protection zones in Cumbria and Dumfries and Galloway; slaughtering flocks that had received sheep from Longtown, Welshpool and Northampton markets; slaughtering flocks that had received sheep from dealers known to have handled infected flocks.

28 In his submission to the Anderson Inquiry (2002), Professor King explains the rationale for the contiguous cull policy: analysis has shown that some 30% of new cases of FMD were found within 1 km of previously infected premises, and 70–80% within 3 km. This is based on statistical evidence from the recorded cases and does not rely on establishment of a causal link

between cases, since we still do not know the transmission routes of the virus in this outbreak. Nor does it rely on the mathematical modelling. Therefore culling of susceptible animals on contiguous premises removed a significant proportion of cases which had yet to be identified (because the animals were not showing clinical signs).

29 As documented by a number of sources, including a report by Michael McCarthy in *The Independent* (Don't worry. It's under control: Foot-and-mouth crisis, March 13th, 2001). McCarthy writes that the Governments assertion that FMD was under control was 'being pushed to the limit' and that the handling of the outbreak came under particular attack from the Irish Government, who were concerned about the possible spread of FMD to Ireland. McCarthy quotes Eamon O'Cuiv, an Irish agriculture minister, who stated that 'the proof of the pudding is in the eating...not only are we getting fresh outbreaks every day in Britain, but they are very, very dispersed in terms of geography.' McCarthy goes on to quote Mr. Brown as stating: 'we can say with confidence that we have it under control. What it cannot say with confidence is what is already in the national flock and what affect this will have. We are in for a longer haul and we have to acknowledge that. The possibility that another half a million sheep may have to be slaughtered has come about because of the nationwide movement restrictions on livestock. Many of the animals concerned are ewes from the uplands which have been sent to winter quarters on lowland farms and need to return to their primary holdings, either to give birth to their lambs, or because they are on cattle farms whose owners need the pasture for their cattle. To return them all would involve hundreds, if not thousands of vehicle movements. If this is deemed too big a risk it is possible they may have to be killed for animal welfare reasons, even though they are healthy....[the Government] would be buying them out'.

30 In fact, daily cases of new infected premises were into the teens almost every day until April 27th 2001 despite contiguous culling for over a month.

31 Whilst it was theoretically possible to resist a contiguous cull (see Appendix I), this was in practice difficult and farmers were often put under a great deal of pressure by both DEFRA and their neighbours. The Cumbria Inquiry (2002), for example, notes that way that some of the 'contiguous cull' farmers were dealt with was heavy handed. People who wanted to hold onto their sheep (for example, a valued hefted flock) were sometimes seen as 'letting down' those who had cattle and it caused bad feeling between neighbours. Ironically this could be worse where the animals did survive; somehow the people who kept their stock were seen as having 'got away with it'. There were instances of the police becoming involved because farmers would not let DEFRA vets for 'surveillance' visits because they believed that DEFRA vets were more likely to bring infection to their farms. One person volunteered her sheep after livestock on a neighbouring farm contracted the virus in the hope that it would save her cattle from the possibility of infection, only to have her cattle go down with the disease a few days later. There were also numerous inconsistencies; a respondent whose livestock should have been culled under direct contact guidelines (their contract milker had full blown infection in his own few cattle just a short distance away) avoided the cull as the situation was never picked up by

MAFF, and despite them living in a completely devastated area, they and a few neighbours survived. The Cumbria Inquiry also highlights that in some instances, '3 km-zone farms' had not been slaughtered out although corresponding farms in the area had been.

32 This phrase was used frequently during the early days of the 2001 outbreak, Woods (2001), for example, notes that FMD does not have many human health implications. Such pronouncements mirrored similar advice from Europe and North America. For example, Patty Scharko, an extension veterinarian with the University of Kentucky Livestock Disease & Diagnostic Laboratory, writes that (in response to the UK epidemic) 'let me stress that foot and mouth doesn't cause a problem in humans' (cited in Heald, 2001).

33 The research report, together with additional material, is available to download from Cumbria County Council website: http://www.cumbria.gov.uk/ruralmatters/evidence/fmd.asp

34 The concept of lifescape and lifescape change will be explored in Chapter 7.

35 Kai Erikson, personal communication, October 13th 2003.

36 See, for example, the work of Gubrium & Holstein, 1998; Hermanns, 2000; Ricoeur, 1991; Smith & Sparkes, 2002.

Chapter 3 Of Humans and Animals

37 The nature reserve is open to the public on an arranged access basis. The reserve website gives some background to the project: http://www.watchtree.co.uk/index.html

38 Farmers in England and Wales have a higher mortality rate from suicide than the general population, a fact attributed to easy access to firearms in addition to the multiple chronic stresses of farm life (Hall et al., 2004).

39 BSE is a neurological disease which affects cattle. It was first recognized and defined in the United Kingdom in November 1986. Over the next few years the epidemic grew considerably and affected all parts of the country but to different degrees. It reached its peak in 1992, when 36,680 cases were confirmed in Great Britain, and since then has shown a steady decline. BSE occurs in adult animals in both sexes, typically in animals aged five years and more. Affected animals show signs that include; changes in mental state, abnormalities of posture and movement and of sensation. The clinical disease usually lasts for several weeks and it is invariably progressive and fatal. BSE has been a notifiable disease by law since 1988 (DEFRA, 2007).

40 Prions are abnormal forms of protein that attack the central nervous system and then invade the brain, causing dementia. The best-known prion disease is Creutzfeldt-Jakob disease. CJD exists in a number of forms, most of which are very rare. In 1996, a new type of CJD, variant CJD, was reported. Variant CJD appears to affect younger people than the other forms of the disease. The average age of death is 29 years. The current view on variant CJD is that it has resulted from transmission of infection from BSE in cattle to humans via infectivity in food.

41 All cattle have to be tagged within days of their birth with a unique number printed on a plastic tag clipped into the animals ear. This number is then transferred to the animal's 'passport', the bovine equivalent of an ID card.

The passports are issued and administered by the British Cattle Movement Scheme (BCMS) based in Workington which tracks the life, reproduction, movements and death of every bovine in the country. They are used by vets and farmers as a way of identifying each animal when testing for TB for example, and ensuring traceability in the food chain.
42. Holloway (2001) & Philo and Wolch (1998).
43. Harrison & Burgess (1994), Whatmore (1999), Quinn (1993) and Shepard (1996).
44. See also Philo & Wilbert (2000), Elder *et al.* (1998) and Mullin (1999).
45. Kirby (2003) and Blackshall *et al.* (2001).
46. See Blackshall *et al.* (2001), Fielding & Haworth (1999) and Carver & Samson (2004).
47. Known/well looked after.
48. The quality of meat produced from the carcass, ease of butchering, etc.
49. This practice is less likely to happen in the future due to the continued closure of small local abattoirs.
50. Geoff Brown, Cumbrian farmer, rural activist and champion of Herdwick sheep argues that hefted sheep (of which the two most important breeds in Cumbria are Herdwicks and Rough Fell) are a Cumbrian icon and make a key contribution to its cultural landscape (Brown, 2002).
51. In addition, as on-farm killing of livestock for home consumption is now illegal, a dead animal on a farm nowadays is the result of an accident or sickness in other words something bad, worrying, and/or financially important.
52. While there are many parts to the changing role of agriculture, what it is essentially about is the change from a production oriented countryside to a production oriented countryside (including the preservation of agricultural landscapes, managing farmland for wildlife and biological diversity and the prevention of depopulation in remote areas) and, along with it, a move for farmers away from a specialised producer tied in with distant markets to a small local businessman or environmental manager (Marsden, 1999). In particular, in their role as an environmental manager, increased emphasis has been placed on the provision of public goods which may be defined as: 'A public good is a resource from which all may benefit, regardless of whether they have helped provide the good – I can enjoy the parks in my city even if I do not pay municipal taxes. Public goods are also distinguished by the fact that they are non-rival in that one person's use of the good does not diminish its availability to another person.' (Kollock, 1998:188)

Chapter 4 Things/Materials

53. For an explicit use of this approach in relation to FMD see Donaldson *et al.* (2002), for an implicit example see Law (2006).
54. 'What is the state of dwelling in our precarious age?' asks Heidegger (1971:161).
55. In the classical structuralism of Lévi-Strauss, totems are systems of classification (in the cognitive unconscious mind) which determine what an object can become within language and action.

164 *Notes*

56 The role of DEFRA field officer is discussed in more detail in Chapter 6.
57 From an NFU bulletin on restocking August 2001: A bit more information about restocking was discussed at the DEFRA press conference. Over 200 farms have received their FM7s signing off their cleansing and disinfection and DEFRA have already had enquiries about restocking. The restocking protocol has now been received and DEFRA say that during the 21 day period following the issuing of the FM7, farmers will need to produce a plan for restocking for discussion with DEFRA. This will depend on the local disease situation and will involve animals being reintroduced to all parts of the farm which previously held livestock. For the first 28 days the animals will be checked once a week and then blood tested at the end of the 28 days. If these blood tests are clear then the Form A will then be lifted and the farm will revert to the designations relevant to the locality – e.g. Form D restriction if still within a 3 km protection zone, or Infected Area status. The farmer can alternatively opt to remain stock free for four months after which the Form A will also be lifted. DEFRA are allowing restocking from within the Cumbria Infected Area but animals will have to be blood tested before they are moved onto the farm. DEFRA are also urging farmers to consider diseases when restocking particularly TB.
58 DEFRA Biosecurity Advice, 2001: Keep your guard up; Keep your stock away from your neighbour's stock; Keep movements on and off the farm to an absolute minimum; Clean then disinfect everything and everyone who comes onto or moves off your farm; Always wash before disinfecting; Power wash muck off wellies, waterproof clothing, and vehicles – especially wheel arches. Keep the virus out!
59 Captive bolt guns. The gun is held close to the animal's head and a bolt enters the skull propelled by an explosive cartridge. The bolt is withdrawn. Cattle were also pithed or caned: a flexible rod is inserted through the hole into the brain to ensure a quick death.
60 'David Browns' are probably the most common make of tractor to be seen on UK farms.

Chapter 5   Trauma and Traumatic Experience

61 PTSD first appeared as a diagnosis in the Diagnostic and Statistical Manual of the American Psychiatric Association (DSM) in 1980 (DSM-III), with further revisions in DSM-III-R (1987) and DSM-IV (1994). It was first included in the International Classification of Diseases (ICD) system in 1992 (Turnbull, 1998). The history and social construction and cultural production of the PTSD 'diagnosis' has been carefully traced by Allan Young in his book *The Harmony of Illusions*.
62 The Oklahoma City bombing was a terrorist attack on April 19, 1995, in which a US government office complex was destroyed, killing 168 people.
63 However, Tucker *et al.* (2002) do also note that body handlers responding to the Oklahoma City bombings had higher rates of alcohol-use disorders than did direct trauma victims.
64 A portable roofed structure with feeding troughs inside, open at the ends to allow lambs in to feed but low enough to exclude mature ewes.

65 Ewes have a limited breeding season and three weeks represents a full breeding cycle during that season.
66 This is the text of one of more that 650 'Cumbria Updates' sent by email during 2001 to farmers in the North West.
67 An incurable bacterial disease of sheep that alters the absorption capacity of the intestine.
68 An event where the 'experts' listened to the stories of the 'ordinary people' who experienced FMD in 2001.
69 Echoing the work of Yehuda *et al.* (1998) & Alexander & Wells (1991).
70 We would add a note of caution here, long-term stress and its effects may indeed be a serious problem and we believe this should continue to be researched. The point here is about personal and collective resources which were brought into play at the time and in the following months.

Chapter 6 Working on the Frontline

71 Furnham (1997), Arnold *et al.* (1991) and Dobreva-Martinova *et al.* (2002).
72 Arnold *et al.* (1991), Furnham (1997), Piko (2006) and Mullins (1996).
73 Perrewé *et al.* (2002), Karatepe & Sokmen (2006) and Chang & Hancock (2003).
74 Late in the evening of 6 July 1988, leaking gas on the Occidental Piper Alpha North Sea oil rig ignited, causing a devastating blaze which killed 167 of the 226 men on board. Many of the oil workers leapt 100 ft (30 m) into the sea to escape the fire and toxic fumes, despite being told their jump would almost certainly be fatal. It is still the world's worst-ever offshore oil disaster.
75 Portakabin City is the colloquial term used to describe the FMD management centre (MAFF Ops) set up in a Carlisle car park, which became full of temporary offices, some two stories high, to house additional Government and at one stage Army staff. The Portakabins were situated within a few feet of each other, each one linked by overhead electricity and computer cables. Working conditions were reported to be grim.
76 Meichenbaum & Jaremko (1983); Alexander (1993); Dyregrov *et al.* (1996) and Alexander & Klein (2001).
77 Long *et al.* (2007:306) suggest that the apparent lack of distress amongst 9/11 Red Cross workers may be because the Red Cross workers are a 'psychologically healthy population that may be resistant to stress and distress responses.'
78 Small track or lane.
79 Legal proceedings were pending.
80 Data kindly provided by DEFRA Occupational Health Department in 2002. Information regarding response rates and sampling strategy is not available. This study is highlighted in order to further contextualise the health problems experienced by DEFRA frontline staff.
81 See for example, Alexander & Klien (2001) Bisson *et al.* (2003), Carlier *et al.* (1998), Duckworth (1991), Everly *et al.* (1999) and Jonsson & Segesten (2004).
82 Interestingly, in an earlier study of 9/11 Red Cross workers Gaher *et al.* (2006) highlight the links between frontline workers, negative affect, coping motives, and subsequent alcohol problems, particularly among younger

participants who appeared to be at greatest risk. They do also note that this was a 'well-adjusted sample.'
83 By formal debriefing we refer for example to Critical Incident Stress Debriefing (CISD) as part of a Critical Incident Stress Management (CISM) programme, based largely on the debriefing model created by Jeffrey Mitchell (Carlier *et al.*, 1997; Everly *et al.*, 1999).
84 The 1992 Los Angeles riots, also known as the Rodney King riots, was sparked on April 29, 1992 when a mostly white jury acquitted four police officers accused in the videotaped beating of black motorist Rodney King. Thousands of people in Los Angeles (mainly young black and Latino males), joined in a 'race riot', involving mass law-breaking, including looting, arson and murder. Around 50 to 60 people were killed during the riots.

Chapter 7 Exploring the Lifescape

85 Nazarea (1995); Som and McSweeney (1996); Howorth (1999); Howorth and O'Keefe (1999) and Convery (2004).
86 Agnew (1987); Tuan (1977) (1991); Massey (1994); Bender (2001); Brandenburg & Carroll (1995) and Ingold (2000).
87 Pierre Bourdieu is a highly influential 20th century social theorist and habitus is arguably his most significant contribution. It is certainly the one for which he is best known, and is the central concept of his theory of practice. The Latin word habitus refers to a habitual or typical condition, state or appearance, particularly of the body. Bourdieu was influenced by the work of Mauss (1979:101), who argues that these 'habits' do not just vary at an individual level but especially vary between societies, educations, proprieties and fashions...in them we should see the techniques and work of collective and individual practical reason.'
88 All culled farms were allotted a unique case number with a prefix denoting its status IP (Infected Premises) DC (Dangerous Contact) and CP (Contiguous Premises). These prefixes also served to cement new FMD identities.
89 The reopening of livestock auction marts in 2002 was a cause for celebration, but also an indication of how things had changed and would continue to function under new imperatives. (Markets were also closed again during the 2007 outbreak). For example, an agricultural supplies manager tells how he *planned to attend Wigton [auction mart] for an hour or so...however because unknown to me I had to wear wellington boots I was refused entry and missed seeing my customers. Apparently the biosecurity at (another auction) had been below standard and the DEFRA 'top man' had come to Wigton to ensure total compliance with their regulations.*

Chapter 8 Reconfiguring Disasters

90 Bob McCracken, outgoing President of the British Veterinary Association speech to the BVA Congress 29 September 2005, spoke of an 'ever increasingly threatening disease scenario' in which, for example, FMD and Avian Influenza would gain access to the UK.

Bibliography

Agnew, J. (1987) *Place and Politics*. London: Allen & Unwin.
Alexander, D.A. (2005) Early mental health intervention after disasters. *Advances in Psychiatric Treatment*, Vol. 11, pp.12–18.
Alexander, D.A. & Wells, A. (1991) Reactions of Police Officers to Body-Handling after a Major Disaster, A Before and After Comparison. *British Journal of Psychiatry*, 159: 547–555.
Alexander, D.A. & Klein, S. (2001) Ambulance personnel and critical incidents: impact of accidental and emergency work on mental health and emotional well-being. *The British Journal of Psychiatry*, Vol. 178, pp.76–81.
Alexander, D.A. (1993) Stress among police body handlers: a long-term follow-up. *British Journal of Psychiatry*, Vol. 163, pp.806–808.
al-Naser, F. & Everly, G.S. (1999) Prevalence of post-traumatic stress disorder among Kuwaiti firefighters. *International Journal of Emergency Mental Health*, Vol. 1, No. 2, pp.99–101.
Anderson Inquiry (2002) 'Foot and mouth disease 2001: Lessons Learned Inquiry', Chair, Dr Iain Anderson, Cabinet Office.
Anderson, K. (1997) A walk on the wild side: a critical geography of domestication. *Progress in Human Geography*, 21(4): 463–485.
Arnold, J., Roberston, I.T. & Cooper, C.L. (1991) *Work Psychology*. London: Pitman Publishing.
Augé, M. (1998) *A Sense of the Other*. Stanford: Stanford University Press.
Bailey, C. & Convery, I. (2006). *The Health and Social Consequences of the 2005 Carlisle Floods*. Research project funded by Carlisle Churches.
Bailey, C., Convery, I.T., Baxter, J. & Mort, M. (2006). Different public heath geographies of the 2001 foot and mouth disease epidemic: 'citizen' versus 'professional' epidemiology. *Health & Place*, Vol. 12, pp.157–166.
Barnes, D. (1996) An analysis of the grounded theory method and the concept of culture *Qualitative Health Research*. Vol. 6, pp.429–441.
Bearman, P.S. & Stove, L.K. (2000) Becoming a Nazi: models for narrative networks. *Poetics*, 27: 69–90.
Bender, B. (2001) Landscapes on the move. *Journal of Social Archaeology*, Vol. 1, pp.75–89.
Bennett, K., Carroll, T., Lowe, P. & Phillipson, J. (2002) *Coping with Crisis in Cumbria: Consequences of Foot and Mouth Disease*. Centre for Rural Economy Research Report, University of Newcastle Upon Tyne.
Bickerstaff, K., Simmons, P. & Pidgeon, N. (2006) Situating local experience of risk: Peripherality, marginality and place identity in the UK foot and mouth disease crisis. *Geoforum*, 37: 844–858.
Bisson, J.I., Roberts, N. & Macho, G. (2003) The Cardiff traumatic stress initiative: an evidence-based approach to early psychological intervention following traumatic events. *Psychiatric Bulletin*, Vol. 27, pp.145–147.
Blackshall, J., Manley, J. & Rebane, M. (2001) *The Upland Management Handbook*, English Nature, Peterborough.

Blaikie, P., Cannon, T., Davis, I. & Wisner, B. (1994). *At Risk Natural Hazards, People's Vulnerability, and Disasters*. Routledge: London.
Bonanno, G.A. (2004) Loss, Trauma and Human Resilience. *American Psychologist*, Vol. 59, pp.20–28.
Borgman, A. (1984) *Technology and the Character of Contemporary Life*. Chicago: Chicago University Press.
Bourdieu, P. (1977) *Outline of a Theory of Practice*. Cambridge: Cambridge University Press.
Bourdieu, P. (1985) 'The Genesis of the Concepts of Habitus and *Field'*. *Sociocriticism*, Vol. 2, pp.11–24.
Bourdieu, P. (2000) *Pascalian Meditations*. Stanford, CA: Stanford University Press.
Bourdieu, P. (2002) Response to Throop and Murphy. *Anthropological Theory*, Vol. 2, p. 209.
Bracken, P., Giller, J. & Summerfield, D. (1995) Psychological responses to war and atrocity: the limitations of current concepts. *Social Science & Medicine*, Vol. 40, pp.1073–1082.
Brandenburg, A.M. & Carroll, M.S. (1995) Your place, or mine: the effect of place creation on environmental values and landscape meanings. *Society and Natural Resources*, Vol. 8, pp.381–398.
Braun, B. & McCarthy, J. (2005) Hurricane Katrina and abandoned being, Guest Editorial. *Environment & Planning D: Society & Space*, Vol. 23, 795–809.
Breakwell, G.M. & Wood, P. (1995) Diary techniques, in G.M. Breakwell, Hammond, S. & Fife-Schaw, C. (eds). *Research Methods in Psychology*, pp.294–302, London: Sage.
Briggs, H.M. & Briggs, D.M. (1980) *Modern Breeds of Livestock* (Fourth ed.). New York: Macmillan Publishing Co., Inc.
Brondolo, E., Wellington, R., Brady, N., Libby, D. & Brondolo, T.J. (2007) Mechanism and strategies for preventing post-traumatic stress disorder in forensic workers responding to mass fatality incidents. *Journal of Forensic and Legal Medicine*, doi:10.1016/j.jflm.2007.04.007
Brown, G. (2002) 'Herdwick and Rough Fell Sheep in Cumbria'. Submission to the Cumbria Foot & Mouth Disease Inquiry.
Buckle, P., Marsh, G. & Smale, S. (2003) Reframing Risks, Hazards, Disasters and Daily Life: A Report of Research into Local Appreciation of Risks and Threats. *The Australian Journal of Emergency Management*, Vol. 18, No. 2, May 2003.
Buckle, P. (2007) *Building Partnerships for Disaster Risk Reduction and Natural Hazard Risk Management*. World Bank and the United Nations International Strategy for Disaster Reduction.
Burkitt, I. (2002) Technologies of the self. *Journal for the Theory of Social Behaviour*, Vol. 32, pp.219–237.
Burr, V. & Butt, T. (2000) Psychological distress and postmodern thought, in Fee, D. (eds) *Pathology and the Postmodern: Mental Illness as Discourse and Experience*. London: Sage.
Burton, R., Mansfield, L., Schwarz, G., Brown, K. & Convery, I. (2005) Social Capital in Hill Farming. Report for International Centre for the Sustainable Uplands.
Bush, J., Phillimore, P., Pless-Mulloli, T. & Thompson, C. (2005) Carcass Disposal and Siting Controversy: Risk, Dialogue and Confrontation in the 2001 Foot-and-Mouth Outbreak. *Local Environment*, Vol. 10, pp.649–664.

Buttimer, A. (1976) Grasping the dynamism of lifeworld. *Annals of the Association of American Geographers*. Vol. 66, No. 2, pp.277–293.

Callon, M. (1986) Some elements in a sociology of translation: domestication of the scallops and fishermen of St Brieuc Bay. In Law, J. (ed.) *Power, Action and Belief*, 196–233.

Campbell, D. & Lee, R. (2003) 'Carnage By Computer': the blackboard economics of the 2001 foot and mouth epidemic. *Social & Legal Studies*, Vol. 12, pp.425–459.

Cardeña, E., Holen, A., McFarlane, A., Soloman, Z., Wilkinson, C. & Spiegel, D. (1994) A multisite study of acute stress reactions to a disaster, in Widiger, T.A., Frances, A.J., Pincus, H.A., Ross, R., First, M.B., Davis, W. & Kline, M. (eds) *DSM-IV Sourcebook*, pp.377–391. Washington D.C: American Psychiatric Association.

Carlier, I.V.E., Lamberts, R.D., Van Uchelen, A.J. & Gersons, B.P.R. (1998) Disaster-related post-traumatic stress in police officers: a field study of the impact of debriefing. *Stress Medicine*, Vol. 14, pp.143–148.

Carver, S. & Samson, P. (2004) Eee, It's Wild up North! *Ecos*, Vol. 25 (3/4), pp.29–33.

Cockbain, W. (2006) NFU claim victory on HFA. NFU Press Release 14/12/06.

Chang, E. & Hancock, K. (2003) Role stress and role ambiguity in new nursing graduates in Australia. *Nursing and Health Sciences*, Vol. 5, pp.155–163.

Charmaz, K. (1997) 'Identity Dilemmas of Chronically Ill Men', in Strauss, A. & Corbin, J. (eds) *Grounded Theory in Practice*. Thousand Oaks, California: Sage.

Cogent Bulls Catalogue (2002) Cogent Ltd., Aldford, Chester.

Convery, I.T., Bailey, C., Mort, M. & Baxter, J. (2005) Death in the Wrong Place? Emotional Geographies of the UK 2001 Foot and Mouth Disease Epidemic. *Journal of Rural Studies*, Vol. 21, pp.99–109.

Convery, I. (2004) Unpublished PhD thesis. Forests, Communities & Resources in Post-conflict Mozambique, Lancaster University.

Convery, I. (2006) Lifescapes and Governance: The Régulo System in Central Mozambique. *Journal of African Political Economy*, Vol. 109, pp.449–466.

Convery, I. & Dutson, T. (2008) Wild Ennerdale: A Cultural Landscape. *Journal of Rural & Community Development*. Vol. 3, No. 1. Available at: http://www.jrcd.ca/viewissue.php

Coote, A. & Lenaghan, J. (1997) *Citizens Juries: Theory into Practice*. London: IPPR.

CRED. Emergency Events Database. Accessed 20/10/07 Available at: http://www.em-dat.net/who.htm

Csordas, T.J. (1994) *Embodiment and Experience. The Existential Ground of Culture and Self*. Cambridge: Cambridge University Press.

Cumbria Foot & Mouth Disease Inquiry (2002) *Inquiry Report*. Cumbria County Council. Available at: http://www.cumbria.gov.uk/elibrary/Content/Internet/538/716/37826163827.pdf

Davey, J.D., Obst, P.L. & Sheehan, M.C. (2000) Work demographics and officers' perceptions of the work environment which add to the prediction of at risk alcohol consumption within an Australian police sample. *Policing: An International Journal of Police Strategies & Management*, Vol. 23, pp.69–81.

Daniels, S. (1994) Grasping the dynamism of lifeworld (communication). *Progress in Human Geography*, Vol. 18, No. 4, pp.501–506.

Deaville, J.A. & Jones, L. (2001) The Health Impact of the Foot and Mouth Situation on People in Wales – The Service Providers Perspective. Institute of Rural Health.

Deaville, J. & Jones, L. (2003) *The impact of the foot and mouth outbreak on mental health and wellbeing in Wales*. A summary report to the National Assembly for Wales by the Institute of Rural Health & University of Glamorgan. Accessed: 02/06/08. Available from: http://www.wales.nhs.uk/sites3/Documents/522/footandmouth.pdf

de Beauvoir, S. (1972) *The Second Sex*. Penguin.

DEFRA (2007) Bovine Spongiform Encephalopathy (BSE). Accessed 02/06/08. Available from: http://www.defra.gov.uk/animalh/bse/index.html

Department of Natural Resources and the Environment (2000) Investigatory Committee to the Parliament of Victoria Australia. Inquiry into control of Ovine Johne's disease (OJD) in Victoria. Chapter 11: Social impacts.

Derret, R. (2003) Making sense of how festivals demonstrate a community's sense of place. *Event Management*, Vol. 8, pp.49–58.

DeSilvey, C. (2006) Observed Decay: Telling Stories with Mutable Things. *Journal of Material Culture*, Vol. 11, pp.318–338.

Dick, P.K. (1968) *Do Androids Dream of Electric Sheep?* Gollancz: London.

Dobreva-Martinova, T., Villeneuve, M. Strickland, L. & Matheson, K. (2002) Occupational Role Stress in the Canadian Forces: Its Association With Individual and Organisational Well-Being. *Canadian Journal of Behavioural Science*, Vol. 34, pp.111–121.

Donaldson, A.P., Lowe, P. & Ward, N. (2002) Virus-Crisis-Institutional Change: The Foot and Mouth Actor Network and the Governance of Rural Affairs in the UK, *Sociologica Ruralis*, Vol. 42, No. 3, pp.201–214.

Dowswell, T., Harrison, S., Mort, M. & Lilford, R. (1997) *Health Panels – A Survey*. Final report of a research project funded by NHS Executive (Northern & Yorkshire) Project No HSR 015. Nuffield Institute for Health, Leeds University.

Douglas, Mary (1966) *Purity and Danger*. London: Routledge.

Duckworth, D.H. (1991) Managing Psychological Trauma in the Police Service: from the Bradford Fire to the Hillsborough Crush Disaster. *Journal of Social Occupational Medicine*, Vol. 41, pp.171–173.

Dryden, W. & Aveline, M. (1988) *Group Therapy in Britain*. Milton Keynes: Open University Press.

Dyregrov, A. & Mitchell, J.T. (1992) Work with traumatised children – psychological effects and coping strategies. *Journal of Traumatic Stress*, Vol. 5, pp.51–62.

Dyregrov, A., Kristoffersen, J.I. & Gjestad, R. (1996) Voluntary and professional disaster workers: Similarities and differences in reactions. *Journal of Traumatic Stress*, 9: 541–555.

Egoz, S., Bowring, L. & Perkins, H.C. (2001) Tastes in tension: form, function and meaning in New Zealand's farmed landscape. *Landscape and Urban Planning*, Vol. 57, pp.177–196.

Elder, G., Wolch, J. & Emel, J. (1998) Race, place and the bounds of humanity. *Society and Animals*, 6 (2).

Elliot, P. (1997) The Use of Diaries in Sociological Research on Health Experience. *Sociological Research Online*, Vol. 2, No. 2.

Emergency Management Australia (1998) *The Australian Emergency Manual Glossary)*. Emergency Manual Series. Department of Defence, Australia.

Erikson, K. (1976) *In the Wake of the Flood*. London, UK: George Unwin and Allen.
Erikson, K. (1991) Notes on Trauma and Community. *American Imago*, Vol. 48, No. 4, pp.455–472.
Erikson, K. (1994) *A New Species of Trouble*. New York: W.W. Norton & Company.
Erikson, K. (2003) *Reflections on Disaster and Trauma*. Keynote presentation, Voices of Experience Conference, Carlisle Racecourse, October 2003. Available from: http://www.footandmouthstudy.org.uk/?page=47
Erikson, T.H. (1995) *Small Places, Large Issues*. London: Pluto Press.
Eyre, A. & Payne, L. (2006) Property Rights: returning personal possessions after disasters. *Australian Journal of Emergency Management*, Vol. 21, No. 2.
Everly, G.S., Boyle, S.H. & Lating, J.M. (1999) The effectiveness of psychological debriefing with vicarious trauma: a meta-analysis. *Stress Medicine*, Vol. 15, pp.229–233.
Farnell, B. (2000) Getting out of the habitus: an alternative model of dynamically embodied social action. *Journal of the Royal Anthropological Institute*, Vol. 6, pp.397–418.
Ferdman, B. (1990) Literacy and cultural identity. *Harvard Educational Review*, Vol. 60, pp.181–204.
Fielding, A.H. & Haworth, P.F. (1999) *Upland Habitats*. London: Routledge.
Fillmore, K.M. (1990) Occupational drinking subcultures: an exploratory epidemiological study, in Roman, P.M. (ed.) *Alcohol Problem Intervention in the Workplace: Employee Assistance Programs and Strategic Alternatives*. New York: Quorum Books.
Fortun, K. (2001) *Advocacy after Bhopal: Environmentalism, Disaster, and New Global Orders*. Chicago: University of Chicago Press.
Foucault, M. (1988) Technologies of the self, in Martin, L.H., Gutman, H. & Hutton, P.H. (eds) *Technologies of the Self: A Seminar with Michel Foucault*. Cambridge Mass: MIT Press.
Fullerton, C.S., Ursano, R.J. & Wang, L. (2004) Acute stress disorder, posttraumatic stress disorder, and depression in disaster or rescue workers. *American Journal of Psychiatry*, Vol. 161, pp.1370–1376.
Furnham, A. (1997) *The Psychology of Behaviour at Work*. Hove: Psychology Press.
Frankel, D. (2000) Migration and ethnicity in prehistoric Cyprus: Technology as Habitus. *European Journal of Archaeology*. Vol. 3, pp.167–187.
Franklin, S. (2001) Sheepwatching. *Anthropology Today*. Vol. 17, pp.3–9.
Franklin, S. (2007) *Dolly Mixtures the Remaking of Genealogy*. London: Duke University Press.
Franks, J. (2002) Farming after foot and mouth. Centre for Rural Economy Working Paper 66, University of Newcastle upon Tyne.
Franks, J., Lowe, P., Phillipson, J. & Scott, C. (2003) The Impact of foot and mouth disease on farm businesses in Cumbria. *Land Use Policy*, Vol. 20, pp.159–168.
Gaher, R.M., Simons, J.S., Gerard, A., Jacobs, G.A., Meyer, D. & Johnson-Jimenez, E. (2006) Coping motives and trait negative affect: Testing mediation and moderation models of alcohol problems among American Red Cross disaster workers who responded to the September 11, 2001 terrorist attacks. *Addictive Behaviors*, 31: 1319–1330.

Gallagher, E., Ryan, J., Kelly, L., Leforban, Y. & Wooldridge, M. (2002) Estimating the risk of importation of foot-and-mouth disease into Europe. *Veterinary Record*, 150: 769–772.

Garaway, G. (1996) The case-study model: An organizational strategy for cross cultural evaluation. *Evaluation*, Vol. 2, pp.201–211.

Gesler, W. (1992) Therapeutic landscapes: medical geographic research in light of the new cultural geography, *Social Science & Medicine*, Vol. 34, pp.735–746.

Gesler, W. (1993). Therapeutic landscapes – theory and a case study of Epidauros, Greece. *Environment and Planning D*, Vol. 11, pp.171–189.

Gesler, W. (1996) Lourdes: healing in a place of pilgrimage, *Health & Place*, Vol. 2, pp.95–105.

Glaser, B. (1992) *Basics of Grounded Theory Analysis*. Mill Valley CA: Sociological Press.

Graham, C. (2001) *Foot and Mouth: Carlisle: Heart and Soul*. Carlisle: Small Sister for Radio Cumbria.

Gray, J. (1998). Family farms in the Scottish borders: a practical definition by hill sheep farmers. *Journal of Rural Studies*, Vol. 14, pp.341–356.

Gubrium, J. & Holstein, J. (1998) Narrative practice and the coherence of personal stories. *The Sociological Quarterly*, Vol. 39, pp.163–187.

Gurwitsch, A. (1970) Problems of the lifeworld, in Natanson, M. (ed.) *Phenomenology and Social Reality: Essays in Memory of Alfred Schutz*. Martinus Nijhoff: The Hague.

Habermas, J. (1987) *The Theory of Communicative Action: A Critique of Functionalist Reason* (Translated by McCarthy, T.) London: Polity Press.

Harrison, C. & Burgess, J. (1994) Social constructions of nature: a case study of conflicts over the development of Rainham Marshes. Transactions, *Institute of British Geographers*, Vol. 19, pp.291–310.

Hage, J. & Powers, C. (1992) *Post-Industrial Lives: Roles and Relationships in the 21st Century*. London: Sage.

Halfacree, K. (1995) Talking about rurality: social representations of the rural as expressed by residents in six English Parishes. *Journal of Rural Studies*, Vol. 11, pp.1–20.

Hall, M.J., Anthony, N.G., Ursano, R.J., Holloway, H., Fullerton, C. & Casper, J. (2004) Psychological Impact of the Animal-Human Bond in Disaster Preparedness and Response. *Journal of Psychiatric Practice*, Vol. 10, pp.368–374.

Hammond, J. & Brooks, J. (2001) The World Trade Center attack. Helping the helpers: the role of critical incident stress management. *Critical Care*, Vol. 5, No. 6, pp.315–317.

Harada, T. (2000) Space, materials, and the 'social': in the aftermath of a disaster. *Environment and Planning D: Society and Space*, Vol. 18, pp.205–212.

Haraway, D. (1988) Situated knowledges: The science question in feminism and the privilege of partial perspective. *Feminist Studies*, 14(3): 575–599.

Hart, E. & Bond, M. (1995*) Action Research for Health and Social Care: A Guide to Practice*. Buckingham: Oxford University Press.

Harrison, S. & Mort, M. (1998) Which Champions, Which People? Public and User Involvement in Health Care as a Technology of Legitimation. *Social Policy and* Administration, Vol. 32, pp.60–70.

Hayes, N. (2000) *Doing Psychological Research: Gathering and Analysing Data*. Buckingham: Oxford University Press.

Heald, A.D. (2001) *Prevention Key to Keeping Foot and Mouth at Bay*. University of Kentucky, College of Agriculture news releases. Accessed 02/06/08. Available at: http://www.ca.uky.edu/agc/NEWS/2001/Apr/fmd.htm

Hearns, A. & Deeny, P. (2007) The Value of Support for Aid Workers in Complex Emergencies: A Phenomenological Study. *Disaster Manage Response*, Vol. 5, pp.28–35.

Hedlin, M. & Petersson, G. (1998) Survey of ambulance personnel working environment. *Arbetarskyddsstyrelsen*, Solna 7.

Heidegger, M. (1971) *Poetry, Language, Thought*. New York: Harper & Row.

Hermanns, H. (2000) The Coherence of Incoherent Narratives. *Narrative Inquiry*, 10(1): 223–227.

Hetherington, K. & Munro, R. (1997) *Ideas of Difference. Social Spaces and the Labour of Division*. Oxford: Blackwell.

Holloway, L. (2001) Pets and protein: placing domestic livestock on hobbyfarms in England and Wales. *Journal of Rural Studies*, Vol. 17, pp.293–307.

Holstein, J. & Gubrium, J. (2000) Analysing Interpretive Practice, in Denzin, N. & Lincoln, Y. (eds) *Handbook of Qualitative Research*. London: Sage.

Hood, B. & Seedsman, T. (2004) Psychosocial Investigation of Individual and community responses to the experience of Ovine Johne's Disease in rural Victoria. *Australian Journal of Rural Health*, Vol. 12, pp.54–60.

Howorth, C. (1999) *Rebuilding the Local Landscape*. Ashgate: Aldershot.

Howorth, C. & O'Keefe, P. (1999) Farmers do it better: local management of change in Southern Bukina Faso. *Land Degradation and Development*, Vol. 10, pp.93–109.

Health & Safety Executive (2007) Final report on potential breaches of biosecurity at the Pirbright site 2007. Accessed 02/06/08. Available at: http://www.hse.gov.uk/ news/archive/07aug/finalreport.pdf

Humphrey, N. (1995) Histories. *Social Research*, Vol. 62, pp.477–479.

Husserl, E. (1978) The Origin of Geometry, in Luckmann, T. (ed.) *Phenomenology and Sociology*. Harmondsworth, UK: Penguin Books.

Ingold, T. (1992) Culture and the perception of the environment, in Croll, E. & Parkin, D. (eds) *Bush Base: Forest Farm. Culture, Environment and Development*. London:Routledge.

Ingold, T. (2000) *The Perception of the Environment: Essays on Livelihood, Dwelling and Skill*. Routledge: London.

Irwin, A. (2001) *Sociology and the Environment*. Cambridge: Polity Press.

Johnston, R.J., Gregory, D., Pratt, G. & Watts, M. (2000) *Dictionary of Human Geography*. Oxford: Blackwell Publishing Ltd.

Jones, L. (1995) Response to stress is not necessarily pathological. *BMJ*, Vol. 311, pp.509–510.

Jonsson, A. & Segesten, K. (2004) Guilt, shame and need for a container: a study of post-traumatic stress among ambulance personnel. *Accident and Emergency Nursing*, Vol. 12, pp.215–223.

Jowell, T. (2000) Community development and the Government's policy agenda, in Ashton, P. & Hobbs, A. (eds) *Communities Developing for Health*. Health for All Network.

Karatepe, O.M. & Sokmen, A. (2006) The effects of work role and family role variables on psychological and behavioural outcomes of frontline employees. *Tourism Management*. Article in press, available online at www.sciencedirect.com.

Kashefi, E. & Mort, M. (2000) I'll tell you what I want, what I really really want!, *Report of the SW Burnley Citizens' Jury on Health and Social Care*, commissioned by Burnley Primary Care Group, 2000.

Kearns, R.A. & Barnett, R. (1999) Auckland's starship enterprise: placing metaphor in a children's hospital, in Williams, A. (ed.) *Therapeutic Landscapes: The Dynamic Between Place and Wellness*, pp.169–199. New York: University Press of America.

Kearns, R. & Gesler, W. (1998) *Putting Health into Place: Landscape, Identity and Well Being*. Syracuse: Syracuse University Press.

Kirby, K. (2003) What Might a British Forest Landscape Driven by Large Herbivores Look Like? *English Nature Research Report 530*, English Nature, Peterborough.

Klinkenborg, V. (1993) Barnyard Biodiversity. *Audubon*, Vol. 95, pp.78–88.

Kollock, P. (1998) Social Dilemmas: the anatomy of cooperation. *Annual Review of Sociology*, 4, 183–214.

Koniarek, J. & Dudek, B. (2001) Post-traumatic stress disorder and fire fighters attitude to their job. *Medycyna Pracy*, Vol. 52, No. 3, pp.177–183.

Kumagai, Y., Edwards, J. & Matthew S. Carroll, M.S. (2006) Why are natural disasters not 'natural' for victims? *Environmental Impact Assessment Review*, Vol. 26, pp.106–119.

Labov, W. (1997) Some further steps in narrative analysis. *Journal of Narrative and Life History*, 7: 395–415.

Law, J. (2006) Disaster in Agriculture: or foot and mouth mobilities. *Environment and Planning A*, Vol. 38, pp.227–239.

Law, J. & Singleton, V. (2004) 'A Further Species of Trouble? Disaster and Narrative', published by the Centre for Science Studies, Lancaster University, Lancaster UK. Accessed: 10/06/08. Available at: http://www.lancs.ac.uk/fass/sociology/research/resalph.htm

Latour, B. (1987) *Science in Action: How to Follow Scientists and Engineers Through Society*. Milton Keynes: Open University Press.

Latour, B. (1992) Where are the Missing Masses? Sociology of a Few Mundane Artefacts, in Bijker, W. & Law, J. (eds) *Shaping Technology, Building Society: Studies in Sociotechnical Change*. Cambridge, Mass: MIT Press.

Lee, J. (2007) Experiencing landscape: Orkney hill land and farming. *Journal of Rural Studies*, Vol. 23, pp.88–100.

Lee, S. (2006) *When the Levees Broke*. HBO documentary film.

Liao, Shih-Cheng, Lee, Ming-Been, Lee, Yue-Joe, Weng, Tei, Shih, Fu-Yung, Matthew, H.M. (2002) Association of psychological distress with psychological factors in rescue workers within two months after a major earthquake. *Journal of the Formosan Medical Association*, Vol. 101, pp.169–176.

Long, M.E., Meyer, D.L. & Jacobs, G.A. (2007) Psychological distress among American Red Cross disaster workers responding to the terrorist attacks of September 11, 2001. *Psychiatry Research*, 149: 303–308.

López-Ibor, J. (2005) What is a disaster?, in López-Ibor, J. (ed.) *Disasters and Mental Health*. New York: Wiley.

Lovell, T. (2000) Thinking feminism with and against Bourdieu. *Feminist Theory*, Vol. 1, pp.11–32.

Lowe, P., Edwards, S. & Ward, N. (2001) The foot and mouth crisis: issues for public policy and research. Centre for Rural Economy Working Paper 64, University of Newcastle upon Tyne.

Lowe, P., Ratschow, C., Allinson, J. & Laschewski, L. (2003) Government Decision Making Under Crisis: A Comparison of the German and British Responses to BSE and FMD. Centre for Rural Economy Research Report.

Luckmann, T. (1970) On the boundaries of the social world, in Natanson, M. (ed.) *Phenomenology and Social Reality: Essays in Memory of Alfred Schutz.* Martinus Nijhoff: The Hague.

Macnagten, P., & Urry, J. (1998) *Contested Natures.* London: Sage.

MAFF (1999) A New Direction for Agriculture England Rural Development Plan. London: MAFF.

Mallinson, S. (2001) Listening to Respondents: A Study of Standardised Health Measurement, Unpublished PhD thesis, Institute for Public Health Research & Policy, University of Salford, UK.

Mallinson, S. (2002) Listening to respondents: a qualitative assessment of the Short-Form 36 Health Status Questionnaire. *Social Science and Medicine,* Vol. 54, pp.11–21.

Marsden, T. (1999) Rural futures: the consumption countryside and its regulation. *Sociologica Rrualis,* 39(4): 501–520.

Marcus, G. (1986) Contemporary problems of ethnography in the modern world system, in Clifford, J. & Marcus, G.E. (eds) *Writing Culture: The Poetics and Politics of Ethnography.* Berkeley CA: University of California Press.

Massey, D. (1994) *Space, Place and Gender.* Cambridge: Polity Press.

Massey, D. (1993) Power-geometry and a progressive sense of place, in Bird, J., Curtis, B., Putnam, T., Robertson, G. & Tickner, L. (eds) *Mapping the futures.* London: Routledge.

Massey, D. (2002) Globalisation: What does it mean for geography? Keynote lecture presented at the Geographical Association Annual Conference, UMIST, April 2002.

Matthew, H.M. (2002) Association of psychological distress with psychological factors in rescue workers within two months after a major earthquake. *Journal of the Formosan Medical Association,* Vol. 101, pp.169–176.

Mauss, M. (1979) Body techniques, in Mauss, M. (ed.) *Sociology and Psychology.* London: Routledge.

Mayou, R.A., Ehlers, A. & Hobbs, M. (2000) Psychological debriefing for road traffic accident victims. *British Journal of Psychiatry,* Vol. 176, 589–593.

Milbourne, P. (2003) Hunting ruralities: nature, society and culture in 'hunt countries' of England and Wales. *Journal of Rural Studies.* Vol. 19, pp.157–171.

McFarlane, A.C. (1988) Relationship between psychiatric impairment and a natural disaster: the role of distress. *Psychology and Medicine,* Vol. 18, pp.129–139.

McFarlane, A.C. (2000) Posttraumatic stress disorder: a model of the longitudinal course and the role of risk factors. *Journal of Clinical Psychiatry,* 61 Supplement 5, 15–20.

McGrath, J. (1976) Stress and behaviour in organisations, in Dunnette, M. (ed.) *Handbook of Industrial and Organisational Psychology.* Chicago: Rand-McNally.

McLean, I. & Johnes, M. (2000) *Aberfan: Government and Disasters.* Cardiff: Welsh Academic Press.

Merleau-Ponty, M. (1978) in Luckmann, T. (ed.) *Phenomenology and Sociology.* Harmondsworth, UK: Penguin Books.

Merleau-Ponty, M. (1962) *Phenomenology of Perception.* London: Routledge & Kegan Paul.

Meichenbaum, D. & Jaremko, M.E. (1983) *Stress Reduction and Prevention*. New York: Plenum.
Meth, P. (2003) Entries and Omissions: using solicited diaries in geographical research *Area*, Vol. 35, pp.195–205.
Midgley, M. (1983) *Animals and Why They Matter: A Journey Around the Species Barrier*. Penguin: Harmondsworth.
Minh-ha, T. (1994) Other than myself/my other self, in Robertson, G., Mash, M., Tickner, L., Bird, J., Curtis, B. & Putnam, T. (eds) *Travellers Tales*. pp.9–26, London: Routledge.
Moran, D. (2000) *Introduction to Phenomenology*. London: Routledge.
Morgan, C.A., Hill, S., Fox, P., Kingham, P. & Southwick, S.M. (1999) Anniversary Reactions in Gulf War Veterans: A Follow-Up Inquiry 6 Years After the War. *American Journal of Psychiatry*, Vol. 156, pp.1075–1079.
Mort, M., Convery I.T., Bailey, C. & Baxter, J. (2005) Psychosocial effects of the 2001 UK foot and mouth disease epidemic in a rural population: qualitative diary based study, *BMJ*, Vol. 331, pp.1234–1237.
Mort, M., Convery, I., Bailey, C. & Baxter, J. (2004) The Health and Social Consequences of the 2001 Foot & Mouth Disease Epidemic in North Cumbria. Report to the Department of Health (ref: 121/7499). Lancaster University.
Mouzelis, N. (2000) The Subjectivist-Objectivist Divide: Against Transcendence. *Sociology*, Vol. 34, No. 4, pp.741–762.
Mullins, L. (1996) *Management and Organisational Behaviour*. FT: Prentice Hall.
Mullin, M.H. (1999) Mirrors and windows: sociocultural studies of human–animal relationships. *Annual Review of Anthropology*, Vol. 28, pp.201–224.
Murdoch, J. & Pratt, A. (1993) Rural studies, modernism, postmodernism and post rural. *Journal of Rural Studies*, Vol. 9, pp.411–428.
National Audit Office (NAO) (2002) The 2001 Outbreak of Foot and Mouth Disease. Report by the Comptroller And Auditor General HC 939 Session 2001–2002: 21 June 2002. London: the Stationery Office.
National Trust (2002) A complimentary strategy for controlling foot and mouth disease and retaining healthy livestock. Report prepared by the National Trust in conjunction with The National Sheep Association.
Nazarea, V. (1995) Indigenous Agricultural Decision-making Criteria as a Reflection of Socioeconomic Status and Gender in the Philippines, in Warren, M.D., Brokensha, E. & Slikkerveer, L.J. (eds) *Indigenous Knowledge Systems: The Cultural Dimension of Development*. London: Intermediate Technology Press, pp.49–63.
Nazarea, V. (1999) Lenses and Latitudes in Landscapes and Lifescapes, in Nazarea, V. (ed.) *Ethnoecology: Situated Knowledge/Located Lives*, pp.3–20. Tucson: University of Arizona Press.
Nerlich, B., Hamilton, C.A. & Rowe, V. (2002) Conceptualising Foot and Mouth Disease: The Socio-Cultural Role of Metaphors, Frames and Narratives. *metaphoric.de*. Accessed: 02/06/08. Available at: http://www.metaphorik.de/02/nerlich.pdf
Nesmith, E. (2006) Defining 'Disasters' with Implications for Nursing Scholarship and Practice. *Disaster Management & Response*, Vol. 2, pp.59–62.
North, C. & Hong, B.A. (2000) Project CREST: a new model for mental health intervention after a community disaster. *American Journal of Public Health*, Vol. 90, pp.1057–1058.

North, C., Tivis, L., Curtis, M.J., Pfefferbaum, B., Spitznagel, E.L., Cox, J., Nixon, S., Bunch, K.P. & Smith, E.M. (2002) Psychiatric disorders in rescue workers after the Oklahoma City bombing. *The American Journal of Psychiatry*, Vol. 159, pp.857–859.
O'Keefe, P. & Middleton, N. (1998) *Disaster and Development (The Politics of Humanitarian Aid)*. London: Pluto Press.
'Radiation Curbs on Nine Farms' (2007) North West Evening Mail, Barrow, May 11th, http://beta.nwemail.co.uk/1.3299 (accessed 02/06/08).
Paton, D., Smith, L. & Violanti, J. (2000) Disaster response: risk, vulnerability and resilience. *Disaster Prevention and Management*. Vol. 9, pp.173–180.
Peak District Rural Deprivation Forum (2004) *Hard Times*. A research report into hill farming and farming families in the Peak District. Available at: http://www.pdrdf.org/publications/HardTimes-HillFarmingReport-Jan2004.pdf
Perrewé, P.L., Hochwarter, W.A., Rossi, A.M., Wallace, A., Maignan, I., Castro, S.L., Ralston, D.A., Westman, M., Vollmer, G., Tang, M., Wan, P. & Van Deusen, C.A. (2002) Are work stress relationships universal? A nine-region examination of role stressors, general self-efficacy and burnout. *Journal of International Management*, Vol. 8, pp.163–187.
Picado, A., Guitian, F.J. & Pfeiffer, D.U. (2007) Space–time interaction as an indicator of local spread during the 2001 FMD outbreak in the UK. *Preventive Veterinary Medicine*, Vol. 79, pp.3–19.
Piko, B.F. (2006) Burnout, role conflict, job satisfaction and psychological health among Hungarian health care staff: A questionnaire survey. *International Journal of Nursing Studies*, Vol. 43, No. 3, pp.311–318.
Philo, C. (1992) Neglected rural geographies: a review. *Journal of Rural Studies* Vol. 8, pp.193–209.
Philo, C. & Wilbert, C. (2000) Animal spaces, beastly places, in Philo, C. & Wilbert, C. (eds) *Animal Spaces, Beastly Places*. London: Routledge.
Philo, C. & Wolch, J. (1998) Through the geographical looking glass: Space, place and society–animal relations. *Society and Animals*, Vol. 6, pp.103–118.
Popay, J., Thomas, C., Williams, G., Bennett, S., Gatrell, A. & Bostock, L. (2003) A proper place to live: health inequalities, agency and the normative dimensions of space. *Social Science & Medicine*, Vol. 57, pp.55–69.
Potter, G. (2000) For Bourdieu, Against Alexander: Reality and Reduction. *Journal for the Theory of Social Behaviour*, Vol. 30, pp.229–239.
Quinn, M.S. (1993) Corpulent cattle and milk machines: nature, art and the ideal type. *Society and Animals*, Vol. 1, No. 2.
Quinn, N., & Holland, D. (1987) *Cultural Models in Language and Thought*. New York: Cambridge University Press.
Rahu, M. (2003) Health effects of the Chernobyl accident: fears, rumours and the truth. *European Journal of Cancer*, Vol. 39, pp.295–299.
Raphael, B., Meldrum, L. & McFarlane, A.C. (1995) Does debriefing after psychological trauma work? *BMJ*, Vol. 310, pp.1479–1480.
Reckwitz, A. (2002) The status of the material in theories of culture: from social structure to artefacts. *Journal for the Theory of Social Behaviour*, Vol. 32, No. 2, pp.196–217.

Reissman, C. (1993) 'Narrative Analysis', in Huberman, A. & Miles, M. (eds) (2002) *The Qualitative Researcher's Companion*, pp.217–270. Beverly Hills, CA.: Sage.

Report of the Policy Commission on the Future of Farming and Food (2002). (The Curry Report) Farming and Food a Sustainable Future. London: H.M.S.O (Chairman: Sir Donald Curry).

Reybold, L.E. (2002) Pragmatic epistemology: ways of knowing as ways of being. *International Journal of Lifelong Education*, Vol. 21, pp.537–550.

Reynolds, L.A. & Tansey, E.M. (2003) Foot and Mouth Disease: The 1967 Outbreak And Its Aftermath. *Wellcome Witnesses to Twentieth Century Medicine*, Volume 18.

Riba, S. & Reches, H. (2002) When terror is routine: how Israeli nurses cope with multi-casualty terror. Online Journal of Issues in Nursing 2002;7. Accessed: 02/06/08. Available at: http://www.nursingworld.org/MainMenuCategories/ANAMarketplace/ANAPeriodicals/OJIN/TableofContents/Volume72002/Number3September30/IsraeliNursesandTerror.aspx

Ricoeur, P. (1991) *From Text to Action. Essays in Hermeneutics, II*. Illinois: Northwestern University Press.

Richmond, R. (2002) Learners Lives: A Narrative Analysis *The Qualitative Report* (7) 1 March. Accessed: 02/06/08. Available from: http://www.nova.edu/ssss/QR/QR7-3/richmond.html

Risk & Policy Analysts (2005) *The appraisal of human-related intangible impacts of flooding*. Technical Report.

Robin, C. (2002) Outside of houses. *Journal of Social Archaeology*, Vol. 2, pp.245–268.

Robinson, R.C. & Mitchell, J.T. (1993) Evaluation of psychological debriefings. *Journal of Traumatic Stress*, Vol. 6, pp.367–382.

Rose, G. (1993) *Feminism and Geography*. Cambridge: Polity Press.

Rose, G. (1997) *Lifelines: Biology, Freedom, Determinism*. London: Penguin.

Rose, G. (1999) Performing space, in Massey, D., Allen, J. & Sarre, P. (eds) *Human Geography Today*. Cambridge: Polity Press.

Royal Society (2002) Inquiry into Infectious Diseases in Livestock. Plymouth, UK.

Rybarczyk, P., Koba, Y., Rushen, J., Tanida, H. & De Passillé, A.M. (2001) Can cows discriminate people by their faces? *Applied Animal Behaviour Science*, Vol. 74, pp.175–189.

Sayer, A. (2000) System, Lifeworld and Gender: Associational versus Counterfactual Thinking. *Sociology*, Vol. 34, pp.707–725.

Scarpaci, J. (1999) Healing landscapes: revolution and health care in post-socialist Havana, in Williams, A. (ed.) *Therapeutic Landscapes: The Dynamic Between Place and Wellness*, pp.201–220. New York: Oxford University Press.

Schouten, R., Callahan, M.V. & Bryant, S. (2004) Community Response to Disaster: The Role of the Workplace. *Harvard Review of Psychiatry*, Vol. 12, pp.229–237.

Schutz, A. (1940) [1978] Some structures of the lifeworld, in Luckmann, T. (ed.) *Phenomenology and Sociology*. Penguin Books, UK: Harmondsworth.

Schutz, A. (1967) *The Phenomenology of the Social World*. Evanston: Northwestern University Press.

Schutz, A. (1970) *On Phenomenology and social relations*. Chicago: University of Chicago Press.

Shurmer-Smith, P. (2002) *Doing Cultural Geography*. London: Sage Publications.
Seamon, D. (1979) *A Geography of the Lifeworld*. London: Croom Helm.
Secor-Turner, M. & O'Boyle, C. (2006) Nurses and emergency disasters: What is known. *American Journal of Infection Control*, Vol. 34, pp.414–420.
Seeland, K. (1997) *Nature is Culture*. London: IT Publications.
Shamai, S. (1991) Sense of place: An empirical measurement. *Geoforum*, Vol. 22, pp.347–358.
Shepard, P. (1996) *The Others: How Animals Make us Human*. UK: Island Press.
Shepherd, M. & Hodgkinson, P.E. (1990) 'The hidden victims of disaster: helper stress', *Stress Medicine*, Vol. 6, pp.29–35.
Shurmer-Smith, P. & Hannam, K. (1994) *Worlds of Desire, Realms of Power: A Cultural Geography*. London: Hodder Arnold.
Sideris, T. (2003) War, gender and culture: Mozambican woman refugees. *Social Science & Medicine*, Vol. 56, pp.713–724.
Sizoo, E. (1997) *Women's Lifeworlds*. London: Routledge.
Skinner, C. & Mersham, G. (2002) *Disaster Management: A Guide to Issues Management and Crisis Communication*. Oxford: Oxford University Press.
Small, L.M. (2000) Material Memories Ethnography. *Journal of Contemporary Ethnography*, Vol. 29, pp.563–592.
Smith, M. (2002) The 'Ethical' space of the abattoir: on the (in) human(e) slaughter of other animals. *Human Ecology Review*, Vol. 9, pp.49–58.
Smith, B. & Sparkes, A. (2002) Men, sport, spinal cord injury and the construction of coherence: narrative practice in action. *Qualitative Research*, Vol. 2, pp.258–285.
Somé, S. & McSweeney, K. (1996) *Assessing sustainability in Burkina Faso*. ILEIA Newsletter. ETC Leusden, The Netherlands.
Stewart, D. & Mickunas, A. (1974) *Exploring Phenomenology, A Field Guide to the Field and its Literature*. Chicago: American Library Association.
Strauss, A. (1987) *Qualitative Analysis for Social Scientists*. Cambridge: Cambridge University Press.
Strauss, A. & Corbin, J. (1994) Grounded theory methodology: an overview, in Denzin, N. & Lincoln, Y. (eds) *Handbook of Qualitative Research*. Thousand Oaks CA: Sage.
Strauss, A. & Corbin, J. (1998) *Basics of Qualitative Research Techniques and Procedures for Developing Grounded Theory* (2nd edition) Beverly Hills, CA: Sage.
Storper, M.J. & Scott, A.J. (2002) The Geographical Foundations and Social Regulation of Flexible Production Complexes, in Dear, M.J. & Flusty, S. (eds) *Spaces of Modernity*, Oxford: Blackwell.
Summerfield, D. (2001) The invention of post-traumatic stress disorder and the social usefulness of a psychiatric category. *BMJ*, Vol. 322, pp.95–98.
Sutmoller, P. & Casas Olascoaga. R. (2002) Evidence for the Temporary Committee on Foot-and Mouth Disease of the European Parliament, Meeting 2nd September 2002, Strasbourg.
Taylor, A. & Davis, H. (1998) Individual humans as discriminative stimuli for cattle (Bos tsurus). *Applied Animal Behaviour Science*, Vol. 58, pp.13–21.
Teo, P. & Huang, S. (1996) A Sense of Place in Public Housing. A Case Study of Pasir Ris, Singapore. *Habitat International*, Vol. 20, pp.307–325.
Tilley, C. (1999) *Metaphor and Material Culture*. Oxford: Blackwell Publishers.
Thrift, N.J. (1996) *Spatial Formations*. London: Sage.

Throop, C.J. & Murphy, K.M. (2002) Bourdieu and Phenomenology. *Anthropological Theory*, Vol. 2, pp.185–207.
Tuan, Y.F. (1991) A view of geography. *Geographical Review*, Vol. 81, pp.94–107.
Tuan, Y.F. (1977) *Space and Place*. London: Arnold.
Tucker, P., Pfefferbaum, B., Doughty, D.B., Jones, D.E., Jordan, F.B. & Nixon, S.J. (2002) Body handlers after terrorism in Oklahoma City: Predictors of post-traumatic stress and other symptoms. *American Journal of Orthopsychiatry*, Vol. 72, pp.469–475.
Turnbull, G.J. (1998) A review of post-traumatic stress disorder. Part I: Historical development and classification. *Injury*, Vol. 9, pp.87–91.
United Nations Development Programme (2004) *Reducing Disaster Risk: A Challenge for Development*. United Nations Development Programme Bureau for Crisis Prevention and Recovery, New York.
United Nations International Strategy for Disaster Reduction (UN/ISDR) (2004) *Living with Risk A global review of disaster reduction initiatives*. Accessed: 13/03/07 Available from: http://www.unisdr.org/eng/about_isdr/bd-lwr-2004-eng.htm
United Nations Economic Commission for Latin America and the Caribbean (2003) *Handbook for Estimating the Socio-economic and Environmental Effects of Disasters*. Accessed 02/06/08. Available from: http://www.undp.org/cpr/disred/documents/publications/eclac_handbook.pdf
Ursano, R.J., Fullerton, C.S. & McCaughey, B.G. (1994) Trauma and disaster, in Ursano, R.J., McCaughey, B.G. & Fullerton, C.S. (eds) *Individual and Community Responses to Trauma And Disaster: The Structure of Human Chaos*. UK: Cambridge University Press.
Ursano, R.J., Fullerton, C.S., Vance, K. & Kao, T.C. (1999) Posttraumatic stress disorder and identification in disaster workers. *The American Journal of Psychiatry*, Vol. 156, pp.353–359.
Verbrugge, L.M. (1980) Health Diaries. *Medical Care*, Vol. 18, pp.73–95.
Violanti, J.M. (1993) 'What does high stress police training teach recruits? An analysis of coping', *Journal of Criminal Justice*, Vol. 21, pp.411–417.
Violanti, J.M. & Aron, F. (1993) Police Stressors: Variations in Perception Among Police Personnel. *Journal of Criminal Justice*, Vol. 23, pp.287–294.
Violanti, J.M. & Paton, D. (1999) *Police Trauma: Psychological Aftermath of Civilian Combat*. Springfield: Charles C. Thomas.
Wacquant, L.D. (1992) 'Toward a Social Praxeology: The Structure and Logic of Bourdieu's Sociology', in Pierre Bourdieu & Wacquant, L. (eds) *An Invitation to Reflexive Sociology*. Chicago, IL: University of Chicago Press.
Ward, N., Donaldson, A. & Lowe, P. (2004) Policy framing and learning the lessons from the UK's Foot and Mouth Disease crisis. *Environment and Planning C: Government and Policy*, 22: 291–306.
Ward, M., Laffan, S.W. & Highfield, L.D. (2007) The potential role of wild and feral animals as reservoirs of foot-and-mouth disease. *Preventive Veterinary Medicine*, 80, 9–23.
Weisæth, L., Knudsen, Ø. & Tønnessen, A. (2002) Technological disasters, crisis management and leadership stress. *Journal of Hazardous Materials*, Vol. 93, pp.33–45.
Whatmore, S. (1999) Nature: culture, in Cloke, P., Crang, M. & Goodwin, M. (eds) *Introducing Human Geographies*, pp.4–11. London: Arnold.

Williams, A. (1999) *Therapeutic Landscapes: The Dynamic Between Place and Wellness*. UK: Rowman & Littlefield Publishing Group.
Wilson, K. (2003) Therapeutic landscapes and First Nations peoples: an exploration of culture, health and place. *Health & Place*, Vol. 9, pp.83–93.
Wilson, J.P. (1980) Conflict, stress and growth: the effects of the Vietnam war on psychological development of Vietnam veterans, in Figley, C.R. & Leventman, S. (eds) *Strangers at Home: Vietnam Veterans Since the War*. New York: Praeger.
Wittnich, C. & Belanger, M. (2008) How is Animal Welfare Addressed in Canada's Emergency Response Plans? *Journal of Applied Animal Welfare Service*, Vol. 11, pp.125–132.
Wolch, J. & Emel, J. (1998) *Animal Geographies: Place, Politics and Identity in the Nature–Culture Borderlands*. London: Verso.
WHO (1993) *The ICD-10 Classification of Mental and Behavioural Disorders*. Geneva: WHO.
WHO (2002) *Disasters & Emergencies*. Panafrican Emergency Training Centre, Addis Ababa.
WHO (2006) *Health Effects of the Chernobyl Accident and the Special Heath Care Programmes*. Report of the UN Forum Chernobyl Expert Group Health. Geneva: WHO.
Woods, A. (2001) Kill or cure? *The Guardian*. Wednesday February 28th 2001.
Woods, A. (2002) *A manufactured plague? The History of Foot and Mouth Disease in Britain*. London: Earthscan.
Wynne B. (1997) May the Sheep Safely Graze? A Reflexive View of the Expert-Lay Knowledge Divide, in Lash, S., Szerszyinski, B. & Wynne, B. (eds) *Risk, Environment and Modernity: Towards a New Ecology*, pp.45–83. London: Sage.
Yarwood, R. & Evans, N. (1998) New places for 'old spots': the changing geographies of domesticated livestock. *Society and Animals*, 6 (2).
Yarwood, R. & Evans, N. (2000) Taking stock of farm animals and rurality, in Philo, C. & Wilbert, C. (eds) *Animal Spaces, Beastly Places*. London: Routledge.
Yehuda, R., McFarlane, A.C. & Shalev, A.Y. (1998) Predicting the development of PTSD from the Acute Response to a Traumatic Event. *Biological Psychiatry*, 44, 1305–1313.
Young, A. (1995) *The Harmony of Illusions: Inventing Post Traumatic Stress Disorder*. Princeton: Princeton University Press.
Zimmerman, D.H. & Wieder, D. (1977) The Diary-Interview Method. *Urban Life*, Vol. 5, pp.479–498.

Index

Aberfan, 2, 86, 154
Anderson Inquiry, 18–19, 23, 72
Argentina (FMD), 15
Army, 30–31, 145 *see also* military
Artificial insemination, 49
Animal suffering, 17, 22, 120

Biosecurity, 21, 72, 79, 101, 122
Bhopal, 9
Buffalo Creek, 9, 85, 154
Business
   Change, 46, 49, 103–105
   Loss of, 22, 34, 78, 87–88, 95, 103–105, 107, 146

Cattle
   Beef, 17, 28, 46, 48
   Dairy, 17, 28, 46, 48, 56–57, 121
   Post fmd problems, 102, 143
Citizen panel, 34–35
Clean/dirty, 40, 73, 91, 105, 108, 116
Chernobyl, 8
Children
   Effect on, 79, 81–82, 96–97, 105, 112, 139, 147, 149
Co-ethnography, 35–38
Communication, 10, 46, 47, 74, 79, 99, 117, 122
Cultural landscape, 51, 54
Cumbria Inquiry, 23, 28, 31, 92, 95, 159–161
Culls
   Biosecurity, 122
   Compulsory, 155
   Contiguous, 29, 153
   Decomposition, 77
   Future management, 107
   Government policy, 28–31
   Healthy stock, 34, 41, 106
   Scale, 23
   Threat, 46
   Trauma, 16, 48, 51–52, 57, 106, 119

Diary research, 35
Debriefing, 127
Disaster
   Defining, 5, 151
   Ethnography, 10
   History, 8
   Plot, xvi
   Toxic, xii, xv, 9
Disposal, 23, 29, 32, 43, 68, 72, 74, 76–77, 92, 95, 119, 149
   see also culls
DEFRA, 27, 47, 57, 72–73, 86, 101, 118, 126, 130, 145, 155
Distance/marginality, 85, 98

Economy (rural), 22, 143, 152
Ethnoplot, 36
Epidemiological modelling, 29

Family
   Lost time, 37, 96, 97, 143, 145
   Relationships, 105, 117, 147
Farming lifescape, 43–49, 52, 57
Field Officer, 114–116
Foot and mouth disease
   Description, 14
   Economic impacts, 23
   Presentation, 16
Frustration, 32, 47, 101, 104, 118, 143, 145, 149

Governance (rural), 59, 85, 144
Great Orton, 41 *see also* Watch Tree

Habitus, 54, 135
Health, 33, 38, 90, 92, 115, 117, 123, 133, 142, 147, 148, 149
Health & Safety Executive, 21
Heft, 48, 53–54
Hope/comfort, 96, 100, 108, 146

Identity, 50, 51–52, 56–57, 96, 112, 133, 142, 147, 150, 152
Income
  Higher earnings, 69, 122
  Loss of, 22, 49, 78, 99, 103
Insider/outsider, 104, 109, 140, 149
Isolation, 32, 69, 99, 101, 103, 139

Knowledge
  Distal, 85, 153
  Local, 29–30, 44, 75, 85, 153
  Professional, 29, 85, 87
  Proximal, 85, 153
Kobe earthquake, 2

Lifescape
  Defining, 132
  Discourse/terminology, 47, 132–134
  Disruption, 132, 138, 143
  Farming, 43–49, 52, 57
Life's work, 57–58, 93
Lifeworld, 135
Livestock (on farm), 48–49
  Auctions, 20, 27–28, 44–45, 50, 92, 140, 142, 145
  Bloodlines, 48, 57–58, 102, 105, 143
  Transport, 20, 27–28, 47, 66–67, 95, 101
Longtown, 27–28, 75–76, 128
Loss, 13, 52, 54, 59, 90, 93, 106, 112, 142

MAFF, 27
Mass Observation Movement, 11
Military, 30 *see also* army

New Orleans, 10–11, 86, 154
New species of trouble, 9, 151
Normality, 47, 73, 89, 142, 147

Occupational stress, 115
Occupation change, 115
Ojibwa, 62
Ovine Johnes Disease, 106

Peer support, 127
Phenomenology, 136–137

Picturesque, 43
Pirbright, 19–21
Planning/contingency, 18–20, 85, 114, 131, 154
PTSD, 89–91, 113, 125
Pyres, 3, 27, 33, 43, 48, 74, 148, 149

Radio Cumbria, 13, 106
Relationships
  see also family
Restocking, 48, 69, 101–102, 143, 164
Role
  Ambiguity, 115
  Conflict, 115
Romantic (view of) countryside, 43

School, 96–97
Sense of place, 111, 134
Sheep, 8, 20, 28, 30, 34, 42, 43, 48–49, 51, 53–55, 57, 81, 101, 113, 139, 155
Slaughtermen, 16, 31, 52, 57, 91, 95, 102, 114, 121–122
Snowie, 68–69
Social practices, 63, 81, 98–99
Spaces (public/private), 46, 64, 69, 73, 146
Stigma, 52, 64, 70, 73, 105, 109, 147
Stress, 89–92, 108, 112, 114–116, 118, 122–123, 126, 128, 139, 147

Therapeutic spaces, 130
Totemic, 63–64
Tourism, 22, 24, 103
Traditions, 46, 50, 54, 98, 100, 110
Trauma
  Collective/individual, xiv, 92, 105, 110
  Debriefing, 127
  Defining, 89
  Disaster worker, 116–119, 125
  Public, 147
  Shared, 109
  Reconfiguring, 112–113, 152–153
  Sensory (smell), 16, 52, 75, 77, 100, 113, 119–120, 149
Traumatic experience, 90, 112

Trust
  At local level, 59, 106, 129, 130
  In governance, 59, 85, 101, 138, 146–147, 149, 151, 152

Uruguay (FMD), 17

Vaccination, 18, 159

Vets, 16–17, 19, 23, 28, 46, 69, 98–99, 100, 101, 107, 118, 122, 145

War, 104–106, 108–109, 114, 124
Watchtree Nature Reserve, 41
  see also Great Orton

Zoonosis, 33